William Edward Shuckard

British Bees

An introduction to the study of the natural history and economy of the bees

indigenous to the British Isles

William Edward Shuckard

British Bees

An introduction to the study of the natural history and economy of the bees indigenous to the British Isles

ISBN/EAN: 9783337412937

Printed in Europe, USA, Canada, Australia, Japan

Cover: Foto ©Andreas Hilbeck / pixelio.de

More available books at **www.hansebooks.com**

BRITISH BEES:

AN INTRODUCTION

TO THE STUDY OF THE

NATURAL HISTORY AND ECONOMY OF THE BEES

Indigenous to the British Isles.

BY

W. E. SHUCKARD.

AUTHOR OF 'ESSAY ON THE FOSSORIAL HYMENOPTERA,' 'COLEOPTERA DELINEATED,' 'ELEMENTS OF BRITISH ENTOMOLOGY,' MONOGRAPHS OF THE 'DORYLIDÆ,' 'AULACIDÆ,' ETC. ETC.; AND TRANSLATOR OF BURMEISTER'S 'MANUAL OF ENTOMOLOGY.'

LONDON:
LOVELL REEVE & CO., 5, HENRIETTA STREET, COVENT GARDEN.
1866.

TO

WILLIAM WILSON SAUNDERS, ESQ.,

F.R.S., TREAS. & V.P.L.S., F.Z.S.,

TREASURER OF THE ROYAL HORTICULTURAL SOCIETY,

ETC. ETC. ETC.,

IN TESTIMONY OF THE ABILITY, ZEAL, AND LIBERALITY

WITH WHICH HE CULTIVATES AND PROMOTES

THE SCIENCE OF ENTOMOLOGY;

AND

AS AN ACKNOWLEDGMENT OF MUCH KINDNESS,

EXTENDING OVER MANY YEARS,

This Volume

IS RESPECTFULLY INSCRIBED,

BY HIS FAITHFUL SERVANT,

W. E. SHUCKARD.

PREFACE.

A FEW words are necessary explanatory of the course pursued in the following work, as regards the citation of authorities.

All the facts recorded without reference to authorities, are the result either of personal observation or of diligent study, which, from the length of time that has intervened, have become so blended in my mind that I can no longer separate their sources. I may, however, state that observation has, certainly, as often anticipated the perusal of the discoveries of others, as their record has stimulated direct observation to confirm them.

The habits of animals, in which instinct is the sole prompter, are so uniform, that these, once well observed, may be considered as permanently established. The slight deviations that have been occasionally noticed, although temporarily infringing, do not abrogate the inflexibility of the law which regulates this faculty; and the descendants inevitably resume the economy of the ancestor.

The merit that attaches to the discovery of such facts is due merely to patience and diligence, very common attributes; and the repeated mention of the supposed first observer must, necessarily, in a work of this kind, which is far from being of a strictly scientific character, diminish the interest of the narrative by interrupting its connection, and thus making it an incongruous mosaic. The omission to cite authorities may also take place without any wish to detract from the merit of the discoverer, which is patent to all by his own record in the archives of science.

Before concluding, I wish to express my best thanks to Thomas Desvignes, Esq., for the kindness and willingness with which he lent me, for the purposes of this work, my own selection from the Bees of his choice collection of British insects.

I now dismiss the book—truly a labour of love—with the hope that it will fall into the possession of many, who may be sufficiently interested in the subject to induce them to become ardent entomologists, by showing them within how small a compass much agreeable instruction lies.

June, 1866.

CONTENTS.

CHAPTER I.

PRELIMINARY OBSERVATIONS, COMPRISING GENERAL REMARKS UPON THE USES OF BEES IN THE ECONOMY OF NATURE, THEIR DIVISION INTO SOCIAL AND SOLITARY, AND A NOTICE OF THEIR FAVOURITE PLANTS . . . 1

CHAPTER II.

GENERAL HISTORY OF BEES 17
THE EGG 18
THE LARVA 19
THE PUPA 22
THE IMAGO 23

CHAPTER III.

SKETCH OF THE GEOGRAPHY OF THE GENERA OF BRITISH BEES 61

CHAPTER IV.

NOTICE OF THE MORE CONSPICUOUS FOREIGN GENERA . . 101

CHAPTER V.

PARASITES OF BEES AND THEIR ENEMIES 109

CHAPTER VI.

GENERAL PRINCIPLES OF SCIENTIFIC ARRANGEMENT . . 118

CHAPTER VII.

BRIEF NOTICE OF THE SCIENTIFIC CULTIVATION OF BRITISH BEES 142

CHAPTER VIII.

A NEW ARRANGEMENT OF BRITISH BEES, WITH ITS RATIONALE, AND AN INTRODUCTION TO THE FAMILY, SUBFAMILIES, SECTIONS, AND SUBSECTIONS 153

CHAPTER IX.

A TABLE, EXHIBITING A METHOD OF DETERMINING THE GENERA OF BRITISH BEES WITH FACILITY 170

EASY DISTRIBUTION OF THE BEES 176

CHAPTER X.

THE SCIENTIFIC ARRANGEMENT AND DESCRIPTION OF THE GENERA, WITH LISTS OF OUR NATIVE SPECIES, AND AN ACCOUNT OF THE HABITS AND ECONOMY OF THE INSECTS, WITH INCIDENTAL OBSERVATIONS SUGGESTED BY THE SUBJECT 184

ANDRENIDÆ (SUBNORMAL BEES) 185
   GEN. 1. COLLETES 185
   GEN. 2. PROSOPIS . . . 191
   GEN. 3. SPHECODES . . 196
   GEN. 4. ANDRENA 200
   GEN. 5. CILISSA 211
   GEN. 6. HALICTUS 214
   GEN. 7. MACROPIS 220
   GEN. 8. DASYPODA 224

CONTENTS.

	PAGE
APIDÆ (NORMAL BEES)	227
SCOPULIPEDES (BRUSH-LEGGED BEES)	227
GEN. 9. PANURGUS	227
GEN. 10. EUCERA	231
GEN. 11. ANTHOPHORA	236
GEN. 12. SAROPODA	242
GEN. 13. CERATINA	245
NUDIPEDES (CUCKOO BEES)	249
GEN. 14. NOMADA	249
GEN. 15. MELECTA	255
GEN. 16. EPEOLUS	258
GEN. 17. STELIS	262
GEN. 18. CŒLIOXYS	265
DASYGASTERS (ARTISAN BEES)	269
GEN. 19. MEGACHILE	269
GEN. 20. ANTHIDIUM	279
GEN. 21. CHELOSTOMA	283
GEN. 22. HERIADES	288
GEN. 23. ANTHOCOPA	290
GEN. 24. OSMIA	294
CŒNOBITES (SOCIAL BEES)	302
GEN. 25. APATHUS	302
GEN. 26. BOMBUS	307
GEN. 27. APIS	318
INDEX	363

LIST OF PLATES.

Note.— ♂ signifies male; ♀, female; ☿, neuter.

Plate I.

1. Colletes Daviesiana, ♂ ♀.
2 ♂. Prosopis dilatata.
2 ♀. Prosopis signata.
3. Sphecodes gibbus, ♂ ♀.

Plate II.

1. Andrena fulva, ♂ ♀.
2. Andrena cineraria, ♂ ♀.
3. Andrena nitida, ♂ ♀.

Plate III.

1. Andrena Rosæ, ♂ ♀.
2. Andrena longipes, ♂ ♀.
3. Andrena cingulata, ♂ ♀.

Plate IV.

1. Halictus xanthopus, ♂ ♀.
2. Halictus flavipes, ♂ ♀.
3. Halictus minutissimus, ♂ ♀.

Plate V.

1. Cilissa tricincta, ♂ ♀.
2. Macropis labiata, ♂ ♀.
3. Dasypoda hirtipes, ♂ ♀.

Plate VI.

1. Panurgus Banksianus, ♂ ♀.
2. Eucera longicornis, ♂ ♀.
3. Anthophora retusa, ♂ ♀.

Plate VII.

1. Anthophora furcata, ♂ ♀.
2. Saropoda bimaculata, ♂ ♀.
3. Ceratina cærulea, ♂ ♀.

Plate VIII.

1. Nomada Goodeniana, ♂ ♀.
2. Nomada Lathburiana, ♂ ♀.
3. Nomada sexfasciata, ♂ ♀.

List of Plates

Plate IX.
1. Nomada signata, ♂ ♀.
2. Nomada Fabriciana, ♂ ♀.
3. Nomada flavoguttata, ♂ ♀.

Plate X.
1. Nomada Jacobææ, ♂ ♀.
2. Nomada Solidaginis, ♂ ♀ (that marked ♂ * should be ♀).
3. Nomada lateralis, ♂ ♀.

Plate XI.
1. Melecta punctata, ♂ ♀.
2. Epeolus variegatus, ♂ ♀.
3. Stelis phæoptera, ♂ ♀.

Plate XII.
1. Cœlioxys Vectis, ♂ ♀.
2. Megachile maritima, ♂ ♀.
3. Megachile argentata, ♂ ♀.

Plate XIII.
1. Anthidium manicatum, ♂ ♀.
2. Chelostoma florisomne, ♂ ♀.
3. Heriades truncorum, ♂ ♀.

Plate XIV.
1. Osmia bicolor, ♂ ♀.
2. Anthocopa Papaveris, ♂ ♀.
3. Osmia leucomelana, ♂ ♀.

Plate XV.
1. Apathus rupestris, ♂ ♀.
2 ♂. Apathus campestris (the sexual sign to this should be ♀).
2 ♀. Apathus vestalis.
3. Bombus fragrans, ♀.
4. Bombus Soroensis, ♂ (var. Burrellanus).

Plate XVI.
1. Bombus Harrisellus, ♀.
2. Bombus Lapponicus, ♀.
3. Bombus sylvarum, ♀.
4. Apis mellifica, ♂ ♀ ♀.

BRITISH BEES.

(*HYMENOPTERA.*)

CHAPTER I.

PRELIMINARY OBSERVATIONS,

COMPRISING GENERAL REMARKS UPON THE USES OF BEES IN THE ECONOMY OF NATURE; THEIR DIVISION INTO SOCIAL AND SOLITARY; AND A NOTICE OF THEIR FAVOURITE PLANTS.

It is very natural that the "Bee" should interest the majority of us, so many agreeable and attractive associations being connected with the name. It is immediately suggestive of spring, sunshine, and flowers,—meadows gaily enamelled, green lanes, thymy downs, and fragrant heaths. It speaks of industry, forethought, and competence,—of well-ordered government, and of due but not degrading subordination. The economy of the hive has been compared by our great poet to the polity of a populous kingdom under monarchical government. He says:—

"Therefore doth Heaven divide
The state of man in divers functions,
Setting endeavour in continual motion;

To which is fixed, as an aim or butt,
Obedience: for so work the honey bees;
Creatures, that, by a rule in nature, teach
The act of order to a peopled kingdom.
They have a king, and officers of sorts:
Where some, like magistrates, correct at home;
Others, like merchants, venture trade abroad;
Others, like soldiers, armed in their stings,
Make boot upon the summer's velvet buds;
Which pillage they, with merry march, bring home
To the tent-royal of their emperor:
Who, busied in his majesty, surveys
The singing masons building roofs of gold;
The civil citizens kneading up the honey;
The poor mechanick porters crowding in
Their heavy burdens at his narrow gate;
The sad-ey'd justice, with his surly hum,
Delivering o'er to executors pale
The lazy yawning drone."—*Henry V.*, 1, 2.

Nothing escaped the wonderful vision of this "myriad-minded" man, and its pertinent application.

This description, although certainly not technically accurate, is a superb broad sketch, and shows how well he was acquainted with the natural history and habits of the domestic bee.

The curiosity bees have attracted from time immemorial, and the wonders of their economy elicited by the observation and study of modern investigators, is but a grateful return for the benefits derived to man from their persevering assiduity and skill. It is the just homage of reason to perfect instinct running closely parallel to its own wonderful attributes. Indeed, so complex are many of the operations of this instinct, as to have induced the surmise of a positive affinity to reason, instead of its being a mere analogy, working blindly and without reflection. The felicity of the adaptation of the hexagonal waxen cells, and the skill of the construction of the comb to their purposes, has occupied the abstruse calculations of profound mathematicians; and since human ingenuity has devised modes of investigating, unobserved, the various proceedings of the interior of the hive, wonder has grown still greater, and admiration has reached its climax.

The intimate connection of "Bees" with nature's elegancies, the Flowers, is an association which links them agreeably to our regard, for each suggests the other; their vivacity and music giving animation and variety to what might otherwise pall by beautiful but inanimate attractions. When we combine with this the services bees perform in their eager pursuits, our admiration extends beyond them to their Great Originator, who, by such apparently small means, accomplishes so simply yet completely, a most important object of creation.

That bees were cultivated by man in the earliest conditions of his existence, possibly whilst his yet limited family was still occupying the primitive cradle of the race at Hindoo Koosh, or on the fertile slopes of the Himalayas, or upon the more distant table-land or plateau of Thibet, or in the delicious vales of Cashmere, or wherever it might have been, somewhere widely away to the east of the Caspian Sea,—is a very probable supposition. Accident, furthered by curiosity, would have early led to the discovery of the stores of honey which the assiduity of bees had hoarded;—its agreeable savour would have induced further search, which would have strengthened the possession by keener observation, and have led in due course to the fixing them in his immediate vicinity.

To this remote period, possibly not so early as the discovery of the treasures of the bee, may be assigned also the first domestication of the animals useful to man, many of which are still found in those districts in all their primitive wildness. The discovery and cultivation of the cereal plants will also date from this early age. The domestication of animals has never been satisfactorily explained, but all inquiry seems to point to those regions as the native land, both of them, and of the *gramineæ*, which produce our grain; for Heinzelmann, Linnæus's enthusiastic disciple, found there those grasses still growing wild, which have not been found elsewhere in a natural state.

Thus, long before the three great branches of the human race, the Aryan, Shemitic, and Turonian, took their divergent courses from the procreative nest which was to populate the earth, and which Max Müller proposes to call the Rhematic period, they were already endowed from their patrimony with the best gifts nature could present to them; and they were thus fitted, in their estrangement from their home, with the requirements, which the vicissitudes they might have to contend with in their migrations, most needed. They would eventually have settled into varying conditions, differently modified by time acting conjunctively with climate and position, until, in the lapse of years, and the changes the earth has since undergone, the stamp impressed by these causes, which would have been originally evanescent, became indelible. That but one language was originally theirs, the researches of philology distinctly prove, by finding a language still more ancient than its Aryan, Shemitic, and Turonian derivatives. From this elder language these all spring, their common origin being deduced from the analogies extant in each. These investigations are confirmed by the Scriptural account that "The whole earth was of one language and of one speech," previous to the Flood, and it describes the first migration as coincident with the subsidence of the waters.

That violent cataclysms have since altered the face of the then existing earth, the records of geological science amply show; and that some of mankind, in every portion of the then inhabited world, survived these catastrophes, and subsequently perpetuated the varieties of race, may be inferred from those differences in moral and physical features which now exist, and which have sometimes suggested the impossibility of a collective derivation from one stock. The philological thread, although generally a mere filament of extreme tenuity, holds all firmly together.

That animals had been domesticated in a very early stage of man's existence, we have distinct proof in many recent geological discoveries, and all these discoveries show the same animals to have been in every instance subjugated; thus pointing to a primitive and earlier domestication in the regions where both were originally produced. That pasture land was provided for the sustenance of these animals, they being chiefly herbivorous, is a necessary conclusion. Thence ensues the fair deduction that *phanerogamous*, or flower-bearing plants coexisted, and bees, consequently, necessarily too,—thus participating reciprocal advantages, they receiving from these plants sustenance, and giving them fertility.

These islands, under certain modifications, were, previous to the glacial period, one land with the continent of Europe; and it was when thus connected that those many tropical forms of animal life, whose fossil remains are found embedded in our soil, passed hither. By the comparatively rapid intervention of geological changes, some of the lower forms of life went no further than the first land they reached, and are, consequently, not even now to be found so far west as Ireland: the migration appears clearly to have come from the East. Thus, although we have no direct evidence of the presence of "bees," yet as insects must have existed here, from the certainty that the remains of insect-feeding reptiles are found, as well as those of herbivorous animals, it may be concluded that "bees" also abounded.

Claiming thus this very high antiquity for man's nutritive "bee," which was of far earlier utility to him than the silkworm, whose labours demanded a very advanced condition of skill and civilization to be made available; it is perfectly consistent, and indeed needful, to claim the simultaneous existence of all the bee's allies. The earliest Shemitic and Aryan records, the Book of Job, the Vedas, Egyptian sculptures and papyri, as well as the poems of Homer, confirm the early cultivation of bees by man for domestic uses; and their frequent representation in Egyptian hieroglyphics, wherein the bee occurs as the symbol of royalty, clearly shows that their economy, with a monarch at its head, was known; a hive, too, being figured, as Sir Gardner Wilkinson tells us, upon a very ancient tomb at Thebes, is early evidence of its domestication there, and how early, even historically, it was brought under the special dominion of mankind. To these particulars I shall have occasion to refer more fully when the course of my narrative brings me to treat of the geographical distribution of the "honey bee;" I adduce it now merely to intimate how very early, even in the present condition of the earth, bees were beneficial to mankind, and that, therefore, the connection may have subsisted, as I have previously urged, in the remotest and very primitive ages of the existence of man; and that imperatively with them, the entire family of which they form a unit only, was also created.

In America, where *Apis mellifica* is of European introduction, swarms of this bee, escaping domestication, resume their natural condition, and have pressed forward far into the uncleared wild; and widely in advance of the conquering colonist, they have taken their abode in the primitive, unreclaimed forest. Nor do they remain stationary, but on, still on, with every successive year, spreading in every direction; and thus surely indicating to the aboriginal red-man the certain, if even slow, approach of civilization, and the consequent necessity of his own protective retreat:—a strong instance of the distributive processes of nature. It clearly shows how the wild bees may have similarly migrated in all directions from the centre of their origin. That they are now found at the very *ultima Thule*, so far away from their assumed incunabula, and with such apparent existing obstructions to their distributive progress, is a proof, had we no other, that the condition of the earth must have been geographically very different at the period of their beginning, and that vast geological changes have, since then, altered its physical features. Where islands now exist, these must then have formed portions of widely sweeping continents; and seas have been dry land, which have since swept over the same area, insulating irregular portions by the submergence of irregular intervals, and thus have left them in their present condition, with their then existing inhabitants restricted to the circuit they now occupy. That long periods of time must necessarily have elapsed to have effected this by the methods we still see in operation, is no proof that it has not been. Nature, in her large operations, has no regard for the duration of time. Her courses are so sure that they are ever eventually successful; for, as to her, whose permanency is not computable, it matters not what period the process takes; and she is as indifferent to the seconds of time whereby man's brevity is spanned, as she is to the wastefulness of her own exuberant resources, knowing that neither is lost to the result at which she reaches. Consuming the one, and scattering broadcast the other, but in unnoticeable infinitesimals, she does it irrespective of the origin, the needs, or the duration of man, who can only watch her irrepressible advances by transmitting from generation to generation the record of his observations; marking thus by imaginary stations the course of the incessant stream which carries him upon its surface.

That other bees are found besides the social bees, may be new to some of my readers, who will perhaps now learn, for the first time, that collective similarities of organization and habits associate other insects with "the bee" as bees. Although the names "domestic bee," "honey bee," or "social bee," imply a contradistinction to some other "bee," yet it must have been very long before even the most acute observers could have noticed the peculiarities of structure which constitute other insects "bees," and ally the "wild bees" to the "domestic bee," from the deficiency of artificial means to examine minutely the organization whereby the affinity is clearly proved. This is also further shown in the poverty of our language in vernacular terms to express them distinctively; for even the name of "wild bees," in as far as it has been applied to any except the "honey bee" in a wildered state, is a usage of modern introduction, and of date subsequent to their examination and appreciation. Our native tongue, in the words "bee," "wasp," "fly," and "ant," compasses all those thousands of different winged and unwinged insects, which modern science comprises in the two very extensive Orders in entomology of the *Hymenoptera* and the *Diptera*;—thus exhibiting how very poor common language is in words to note distinctive differences in creatures, even where the differences are so marked, and the habits so dissimilar, as in the several groups constituting these Orders. But progressively extending knowledge, and a more familiar intimacy with insects and their habits, will doubtless, in the course of time, supervene, as old aversions, prejudices, and superstitions wear out, when by the light of instruction we shall gradually arouse to perceive that "His breath has passed that way too;" and that, therefore, they all put forth strong claims to the notice and admiration of man.

It is highly improbable that ordinary language will ever find distinctive names to indicate *genera*, and far less *species:* and although we have some few words which combine large groups, such as " gnats," " flesh-flies," " gad-flies," " gall-flies," " dragon-flies," " sand wasps," " humble bees," etc. etc.; and, although the small group, it is my purpose in the following pages to show in all their attractive peculiarities, has had several vernacular denominations applied to them to indicate their most distinctive characteristics, such as " cuckoo bees," " carpenter bees," " mason bees," " carding bees," etc., yet many which are not thus to be distinguished, will have to wait long for their special appellation.

The first breathings of spring bring forth the bees. Before the hedge-rows and the trees have burst their buds, and expanded their yet delicate green leaves to the strengthening influence of the air, and whilst only here and there the white blossoms of the blackthorn sparkle around, and patches of chickweed spread their bloom in attractive humility on waste bits of ground in corners of fields,—they are abroad. Their hum will be heard in some very favoured sunny nook, where the precocious primrose spreads forth its delicate pale blossom, in the modest confidence of conscious beauty, to catch the eye of the sun, as well as—

> "Daffodils, that come before the swallow dares,
> And take the winds of March with beauty."—*Shakspeare.*

The yellow catkins of the sallow, too, are already swarmed around by bees, the latter being our northern representative of the palm which heralded "peace to earth and goodwill to man." The bees thus announce that the business of the year has begun, and that the lethargy of winter is superseded by energetic activity.

The instinctive impulse of the cares of maternity prompt the wild bees to their early assiduity, urging them to their eager quest of these foremost indicators of the renewed year. The firstling bees are forthwith at their earnest work of collecting honey and pollen, which, kneaded into a paste, are to become both the cradle and the sustenance of their future progeny.

Wherever we investigate wonderful Nature, we observe the most beautiful adaptations and arrangements,—everywhere the correlations of structure with function;

in confirmation of which I may here briefly notice in anticipation, that the bees are divided into two large groups,—the short-tongued and the long-tongued,—and it is the short-tongued,—some of the *Andrenidæ*,—which are the first abroad; the corollæ of the first flowers being shallow and the nectar depositories obvious, an arrangement which facilitates their obtaining with facility the honey already at hand. These bees are also amply furnished,—as will be afterwards explained,—in the clothing of their posterior legs, or otherwise, with the means to convey home the pollen which they vigorously collect, finding it already in superfluous abundance, and which, being borne from flower to flower, impregnates and makes fruitful those plants which require external agents to accomplish their fertility. Thus nature duly provides, by an interchange of offices, for the general good, and by simple, although sometimes obscure means, gives motion and persistency to the wheel within wheel which so exquisitely fulfil her designs, and roll forward, unremittingly, her stupendous fabric.

The way in which the bees execute this object and design of nature, and to which they, more evidently than any other insects, are called to the performance, is shown in the implanted instinct which prompts them to seek flowers, knowing, by means of that instinct, that flowers will furnish them with what is needful both for their own sustenance, and for that of their descendants. Flowers, to this end, are furnished with the requisite attractive qualifications to allure the bees. Whether their odour or their colour be the tempting vehicle, or both conjunctively, it is scarcely possible to say, but that they should hold out special invitation is requisite to the maintenance of their own perpetuity. This, it is supposed, the colour of flowers chiefly effects by being visible from a distance. Flowers, within themselves, indicate to the bees visiting them the presence of nectaria by spots coloured differently from their petals. This nectar, converted by bees into honey, is secreted by glands or glandulous surfaces, seated upon the organs of fructification; and nature has also furnished means to protect these depositories of honey for the bees, from the intrusive action of the rain, which might wash the sweet secretion away. To this end it has clothed the corollæ with a surface of minute hairs, which effectually secures them from its obtrusive action, and thus displays the importance it attaches to the co-operation of the bees. That bees should vary considerably in size, is a further accommodation of nature to promote the fertilization of flowers, which, in some cases, small insects could not accomplish. Many plants could not be perpetuated, but for the agency of insects, and especially of bees; and it is remarkable that it is chiefly those which require the aid of this intervention that have a nectarium, and secrete honey. By thus seeking the honey, and obtaining it in a variety of ways, bees accomplish this great object of nature. It often, also, happens that flowers which even contain within themselves the means of ready fructification cannot derive it from the pollen of their own anthers, but require that the pollen should be conveyed to them from the anthers of younger flowers; in some cases the reverse takes place, as for instance, in the *Euphorbia Cyparissias*, wherein it is the pollen of the older flower which, through the same agency, fertilizes the younger. Although many flowers are night-flowers, yet the very large majority expand during the day; but to meet the requirements of those

which bloom merely at night, nature has provided means
by the many moths which fly only at that time, and
thus accomplish what the bees perform under the eye of
the sun. Here insects are again subservient to the accomplishment of this great act; for the petals of even the
flowers which open in the night only are usually highly
coloured, or where this not the case, they then emit a
powerful odour, both being means to attract the required co-operation. But of course our clients have
nothing to do with these night-blooming flowers, as I
am not aware of a single instance of a night-flying bee;
nor are they on the wing very late in the evening,
being before sunset, already in their nidus. In those
occasional cases where the nectarium of the flower is
not perceptible, if the spur of such a flower which usually becomes the depository of the nectar that has oozed
from the capsules secreting it, be too narrow for the entrance of the bee, and even beyond the reach of its long
tongue, it contrives to attain its object by biting a hole
on the outside, through which it taps the store. The
skill of bees in finding the honey, even when it is much
withdrawn from notice, is a manifest indication of
the prompting instinct which tells them where to seek
it, and is a matter of extreme interest to the observer,
for the honey-marks—the *maculæ indicantes*—surely
guide them; and where these, as in some flowers, are
placed in a circle upon its bosom, as the mark upon
that of Imogen, who had—

> "On her left breast
> A mole cinque-spotted, like the crimson drops
> I' the bottom of a cowslip."—*Shakspeare.*

they work their way around, lapping the nectar as

they go. To facilitate this fecundation of plants, which is Nature's prime object, bees are usually more or less hairy; so that if even they limit themselves to imbibing nectar, they involuntarily fulfil the greater design by conveying the pollen from flower to flower. To many insects, especially flies, some flowers are a fatal attraction, for their viscous secretions often make these insects prisoners, and thus destroy them. To the bees this rarely or never happens, either by reason of their superior strength, or possibly from the instinct which repels them from visiting flowers which exude so clammy a substance. It is probably only to the end of promoting fertilization by the attraction of insects that the structure of those flowers which secrete nectar is exclusively conducive, and which fully and satisfactorily explains the final cause of this organization.

To detect these things, it is requisite to observe nature out of doors,—an occupation which has its own rich reward in the health and cheerfulness its promotes,— and there to watch patiently and attentively. It is only by unremitting perseverance, diligence, and assiduity that we can hope to explore the interesting habits and peculiar industries of these, although small, yet very attractive insects.

Amongst the early blossoming flowers most in request with the bees, and which therefore seem to be great favourites, we find the chickweed (*Alsine media*), the primrose, and the catkins of the sallow; and these in succession are followed by all the flowers of the spring, summer, and autumn. Their greatest favourites would appear to be the *Amentaceæ*, or catkin-bearing shrubs and trees, the willow, hazel, osier, etc., from the male flowers of which they obtain the pollen, and from the female the honey; all the *Rosaceæ*, especially the dog-rose, and *Primulaceæ*, the *Orchideæ*, *Caryophyllaceæ*, *Polygoneæ*, and the balsamic lilies; clover is very attractive to them, as are also tares; and the spots on those leaves of the bean which appear before the flower, and exude a sweet secretion; also the flowers of all the cabbage tribe. Beneath the shade of the lime, when in flower, may be heard above one intense hum of thrifty industry. The blossoms of all the fruit-trees and shrubs, standard or wall, and all aromatic plants are highly agreeable to them, such as lavender, lemon-thyme, mignonette, indeed all the *resedas;* also sage, borage, etc. etc.; but the especial favourites of particular genera and species I shall have occasion subsequently to notice in their series; but to mention separately all the flowers they frequent would be to compile almost a complete flora. Bees are also endowed with an instinct that teaches them to avoid certain plants that might be dangerous to them. Thus, they neither frequent the oleander (*Nerium Oleander*) nor the crown imperial (*Fritillaria imperialis*), and they also avoid the *Ranunculaceæ*, on account of some poisonous property; and although the *Melianthus major* drops with honey, it is not sought. It is a native of the Cape of Good Hope, and may be attractive only to the bees indigenous to the country, which is also the case with other greenhouse plants equally rich in honey, but which not being natives, possibly from that cause the instincts of native insects have no affinity with them.

Bees may be further consorted with flowers by the analogy and parallelism of their stages of existence. Thus, the egg is the equivalent to the seed; the *larva* to the germination and growth; the *pupa* to the bud; and the *imago* to the flower. The flower dies as soon as the seed is fully formed, which is then disseminated by many wonderful contrivances to a propitious soil; and the wild bees die as soon as the store of eggs is as wonderfully deposited, according to their several instincts, in fitting receptacles, and provision furnished to sustain the development of the progeny. Thus, each secures perpetuity to its species, but individually ceases; whereas the unfecundated plant and the celibate insect may, severally, prolong for a short but indefinite period, a brief existence, to terminate in total extinction. Nature thus vindicates her rights, for nothing remains sterile with impunity.

CHAPTER II.

GENERAL HISTORY OF BEES.

THE EGG.—THE LARVA.—THE PUPA.—THE IMAGO.

ALTHOUGH the preceding pages have been written upon the assumption that the reader knows what a bee is, now that we are gradually approaching the more special and technical portion of the subject it will be desirable to conform a little to the ordinary usages of scientific treatment.

The bees constitute a family of the order *Hymenoptera*, viz. insects ordinarily, but in the case of bees always, with four transparent wings, which are variously but partially traversed longitudinally and transversely with threads, called nervures, supposed to be tubular, the relative position of which, together with the areas they enclose, called cells, help to give characters to the genera.

Most of the *Hymenoptera* further possess some kind of an ovipositor,—of course restricted to the females,—varying considerably in the different families. This is sometimes external, but is often seated within the apex of the abdomen, whence it can be protruded for the purpose of depositing the egg in its right nidus. In our insect this organ is converted into a weapon of defence and offence, and forms a sting, supplied by glands with a very virulent poison, which the bee can inject into the wound it inflicts. It is not certain that this organ is used by the bee as an ovipositor, although it is evident it is its analogue. This brief description of the essential peculiarities of the family will, for the present, suffice. In the notice of the imago, I shall enlarge upon the general structure, and then particularize those portions of it which may facilitate further progress.

The Egg.—Although the egg of the parent is the source of the origin of the bee, we cannot abruptly commence from this point, for the preliminary labours of the mother are indispensable to the evolution of its offspring. This egg has to be placed in a suitable depository, together with the requisite food for the sustenance of the vermicule that will be disclosed from it.

Instinct instructs the parent where and how to form the nidus for its egg. These depositories differ considerably in the several genera, but, as a general rule, they are tubes burrowed by the mother either in earth, sand, decaying or soft wood, branches of plants having a pith, the halm of grain, cavities already existing in many substances, and even within the shells of dead snails. These perforations are sometimes simple, and sometimes they have divergent and ramifying channels, Sometimes they are carefully lined with a silky membrane secreted by the insect, and sometimes they are hung with a tapestry of pieces of leaves, cut methodically from plants, but some leave their walls entirely bare. All these particulars I shall have ample opportunity to note in the special descriptions of the genera. I merely indicate them to show how various are the receptacles for the offspring of our bees.

Before the egg is placed within its nidus, this is supplied with the requisite quantity of food needful for the support of the young to the full period of its maturity. The receptacle is then closed, and the same process is repeated again and again until the parent has laid her whole store of eggs. In other cases one tube, or its ramification, contains but one egg. These eggs are usually oblong, slightly curved, and tapering at one extremity; they vary in size according to the species, but are never, however, above a line in length, and sometimes they are very minute. When the stock of the mother bee is exhausted she leaves them to the careful nursing of nature, and the young is speedily evolved. She then wanders forth; time has brought senility; her occupation has gone; and she passes away; but her progeny survive to perpetuate the continual chain of existence.

Fig. 1. The Egg.

The Larva.—The temperature of the perforated tube wherein the egg is deposited must necessarily be higher and more equal than that of the external atmosphere, being secluded from its vicissitudes. The egg is soon hatched, and the larva emerges from its shell to feed ravenously upon the sustenance stored up for its supply. This consists of an admixture of pollen and honey formed into a paste, the quantities varying according to the size of the species. By some species it is formed into little balls; by others, it is heaped irregularly at the bottom of the cell. In the case of *Andrena* the quantity stored is of about the size of a pea. That it must be exceeding nutritious may be inferred from its very nature, consisting, as it does, of the virile, energetic, and fertilizing powder of plants,—the concentration of their living principle. It is strictly analogous to the fecundating property of the semen in animals, and, like them, produces spermatozoa, a fact corroborated by the researches of Robert Brown, Mirbel, and other distinguished vegetable physiologists.*

We are told that the cells of *Hylæus*, or *Prosopis*, and of *Ceratina* are supplied with a semifluid honey. It is very doubtful if *Hylæus* collects its own store, but that *Ceratina* does I have the authority of an exact observer (Mr. Thwaites) to verify it, for he has caught this insect with pollen on its posterior legs, which the long hair covering the tibia is intended for. What may be the nature of this semifluid honey? It is questionable if the larva could be nurtured upon honey alone without the admixture of pollen, thus contradicting analogies presumable from ample verification in nature's processes. How, too, does it become semifluid? It is the property of honey, at a certain temperature, to be very fluid, and this is doubtless the temperature that prevails within the receptacle of the larva during the time of the operations of the bees.

Its semifluid consistency could then apparently be produced only by some more solid admixture, which, if not of pollen, of what can it be? This, even in small quantities, might, upon the bursting of its vesicles, have the power of thickening the fluent honey to the necessary consistency.

But a bee without polliniferous organs cannot collect pollen, and the instance of the hive bee, which collects honey in superabundance, feeding its larva with the bee-

---

* Might not, by parity of inference, the milt of fishes, such as the herring, mackerel, etc., be a useful food in cases of consumption, both from the iodine necessarily existing in it, and also from its doubtless nutritive nature?

bread, must inevitably lead to the conclusion that the larvæ of bees require more than honey for their sustenance. Nature is not usually wantonly wasteful of its resources, and if honey sufficed for the nurture of the grub, so much pollen would not be abstracted from its legitimate purpose, nor would bees have this double trouble given to them. By the admixture of pollen the honey has energetic power infused into it by the spermatozoa which that contains. But it must necessarily be collected, for I never observed, nor have I seen recorded, any instance of the pollen being eaten on the flower and regurgitated into the cell in combination with the imbibed honey.

Pollen is eaten by the domestic bee and humble-bee to form wax for the structure of their cells, but the solitary bees do not themselves consume it.

The larva, when excluded from the egg, is a fleshy

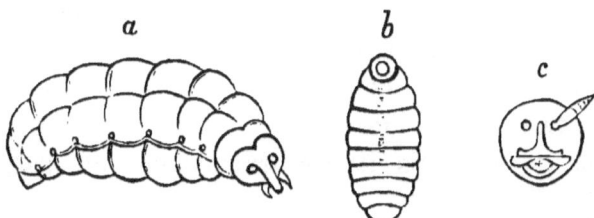

Fig. 2.—*a*, the Larva, when growing; *b*, when preparing to change; *c*, the head, viewed in front.

grub, slightly curved, and a little pointed at each extremity. Its body is transversely constricted, the constrictions corresponding with its fifteen segments, each of which, excepting the head and four terminal ones, is supplied with a spiracle placed at the sides, whereby it breathes; and it has no feet. These segments have on each side a series of small tubercles, which facilitate the restricted motions of the grub, confined to the boundaries of its cell. Its small head, which is smooth above, has a little projecting horn on each side representing the future antennæ. The small lateral jaws articulate beneath a narrow labrum or lip, which folds down over them. To prove that the food provided requires still further comminution, these jaws are incessantly masticating it. The form of these jaws approximates to that of the insect which it will produce, being toothed and broad at the apex in the artisan and wood-boring bees, and simple in those which burrow in softer substances. On each side beneath these jaws there is an appendage, rather plump, having a setiform process at its extremity, and beneath these, in the centre, we observe a fleshy protuberance which, at its tip, has a smaller perforated process that emits the viscid liquid with which the grub spins its cocoon, and which immediately hardens to the consistency of silk.

Having constructed its cocoon, where the species does so,—for it is not incidental to all the genera,—and shrunk to its most compact dimensions, the larva becomes transformed into

The Pupa.—This is semi-transparent at first, and there may be seen through the thin pellicle, which invariably clothes every portion separately, of the body the ripening bee, which lies, like a mummy, with its wings and legs folded lengthwise along its breast. The parts gradually assume consistency, and the natural colours and

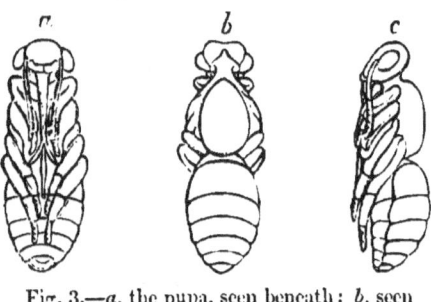

Fig. 3.—*a*, the pupa, seen beneath; *b*, seen above; *c*, seen laterally.

clothing of the perfect insect display themselves through its pellucid envelope. When arrived at perfect maturity, and ready to commence the part it has to perform in the economy of nature, it bursts its cerements, making its way through the dorsal covering of its silken skin, and, leaving the exuviæ behind, it crawls forth from its dormitory, when, becoming invigorated by the bracing air and the genial sunshine, it stretches its legs and expands its wings, and flies forth jubilant, rejoicing in its awakened faculties.

The Imago.—The bee having attained its majority, loses no time in quitting the confined abode wherein it has been hitherto secluded. It comes forth prepared to undertake the cares, and meet the vicissitudes of existence. The new life that now opens to it is one apparently teeming exuberantly with every delight. It dwells in sunshine and amidst flowers; it revels in their sweets, attracted by their beautiful colours and their delightful odours; and the consummation of its bliss is to find a congenial partner. With him it enjoys a brief connubial transport, but which is speedily succeeded by life-long labour, for the cares of maternity immediately supervene.

I believe the wild bees are not polyandrous, and therefore many males, if there be any preponderating discrepancy in favour of that sex, must die celibate. But the fact of finding the males associated together in great numbers upon the same flowers or hedges, is certainly not conclusive of this being the case. To provide a fitting receptacle, furnished with suitable provision, for its future progeny, occupies all the subsequent solicitude of the female.

As frequent reference will hereafter be made to

peculiarities of structure, it will be desirable to take a rapid survey of the external anatomy of the bee, for it will enable me to introduce in due order the requisite technicalities with their local explanations. This course will be found most subservient to preciseness and accuracy, and when mastered, which will be found to be a very simple affair, it will greatly facilitate exact comprehension. No circumlocution can convey what a few technicalities, thoroughly understood, will immediately explain, and no special scientific work can be read with any profit until they are acquired.

Diagrams are introduced to aid the imagination in its conception of what is meant to be conveyed.

This necessary detail I shall endeavour to make as entertaining as I possibly can, by introducing, with the description of the organ, the uses it serves in the economy of the insect. I hope thus to add an interest to it which a merely dry technical and scientific definition would not possess.

Structure is always expressive of the habits of the bees, and is as sure a line of separation, or means of combination, as instinct could be were it tangible. Hence the conclusion always follows with a certainty that such-and-such a form is identical with such-and-such habits, and that, in the broad and most distinguishing features of its economy, the genus is essentially the same in every climate. Climate does not act upon these lower forms of animal life, with the modifying influences it exercises upon the mammalia and man. A *Megachile* is as essentially a *Megachile* in all its characteristics in Arctic America, the Brazils, tropical Africa, Northern China, and Van Diemen's Land, as in these islands, and *Apis* is, wherever it occurs, as truly an *Apis*. Therefore the habits, in whatever country the genus may be found, can thus be as surely affirmed of all its species, from the knowledge we have of those at home, as if observation had industriously tracked them. Therefore, the technicalities of structure once learnt, they become permanently and widely useful.

The body of the bee consists of a head, thorax, and abdomen, which, although to the casual observer, seemingly not separated from each other, are, upon closer inspection, more or less distinctly disconnected. The three parts are merely united by a very short and slight tubular cylinder. This is sometimes so much reduced as to be only a perforation of the parts combined by a ligament, and through which aperture a requisite channel is formed for the passage of the ganglion or nervous chord, which extends from one portion of the body to the other, giving off laterally, in its progress from the sensorium in the head onwards, the filaments required by the organs of sensation and motion, as well as all which control the other functions of the body of the insect.

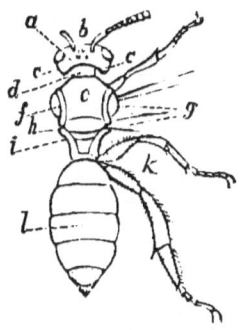

Fig. 1.—Body of the bee. *a*, head and antennæ; *b*, vertex and ocelli; *c*, genæ, or cheeks; *d*, prothorax; *e*, mesothorax; *f*, squamulæ; *g*, insertion of the wings: *h*, scutellum; *i*, post-scutellum; *k*, metathorax; *l*, abdomen.

These apertures form also the necessary medium of connection between the several viscera, whereby the food and other sustaining juices are conveyed from the mouth through the œsophagus to the various parts of the body.

As this work will impinge but very incidentally upon the internal organization of the bee, it is unnecessary to be more explanatory. All that I shall have to notice here are those portions of the external structure which have any special bearing upon the economy and habits, or upon the generic and specific determination of the insects, and to which therefore I shall specially limit myself.

The *head* is the most important segment of the insect's body, if we may elevate to such distinction any portion, when all conduce to the same end, and either would be imperfect without the other, yet we may perhaps thus distinguish it from the rest as it exclusively contains that higher class of organs, those of sense, which are most essential to the functions of the creature. The head consists of the *vertex*, or crown; the *genæ*, or cheeks; the face; the *clypeus*, or nose; the compound eyes; the *stemmata*, or simple eyes; the *antennæ*, or feelers, and the *trophi*, or organs of the mouth collectively.

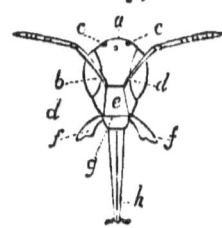

Fig. 5.—Front of the head of the bee. *a*, vertex; *b*, face; *c*, ocelli or stemmata; *d*, compound eyes; *e*, clypeus; *f*, mandibles; *g*, labrum; *h*, lingual apparatus folding for repose.

The *thorax*, the second segment, carries all the organs of locomotion. It consists of the *prothorax* or collar, which carries beneath the anterior pair of legs; the *mesothorax*, or central division, with which articulate laterally above the four wings, the anterior of which have their base protected by the *squamulæ*, or epaulettes, or wing scales, and beneath it carries the intermediate pair of legs; the *metathorax*, or hinder portion, which has in the centre above, behind the *scutellum*, the *post-scutellum*, and at the extremity of this division just above the articulation of the posterior legs is attached the last segment of the insect,—the *Abdomen*.

The *vertex*, or crown of the head, is that portion

which lies between the upper extremities of the compound eyes. Upon the vertex are placed the *stemmata*, or *ocelli* (the simple eyes), in a curve or triangle; they are three in number, and are small, hyaline, circular protuberances, each containing within it a lens; sometimes they occur very far forward upon the face, especially when the compound lateral eyes meet above, as in the male domestic bee or drone. The uses of these simple eyes, from the experiments which have been made, seem to be for long and distant vision. To test their function, Réaumur covered them with a very adhesive varnish, which the bee could not remove, and he then let it escape. He found upon several repeated trials, that the insect always flew perpendicularly upwards, and was lost. Although this was anything but conclusive as to the uses of these eyes, it would seem that by losing the vision of this organ, the insect lost with it all sense of distance.

The *compound eyes*, seated on each side of the head, extend from the vertex generally to the articulation of the mandibles or jaws, their longitudinal axis being perpendicular to the station of the insect. They vary in external shape and convexity in the several species and genera, although not greatly, and consist of a congeries of minute, hexagonal, crystalline facets, each slightly convex externally, and their interstices are sometimes clothed with a short and delicate pubescence. Each separate hexagon has its own apparatus of lens and filament of optic nerve, each having its own distinct vision, but all converge to convey one object to the sensorium. The function of the compound eyes is concluded to be the microscopic sight of near objects.

The *face*, which sometimes has a longitudinal *carina*,

or prominent ridge, down its centre, lies between these eyes, descending from the vertex to the base of the clypeus, or nose, but which is without the function of that organ. This clypeus is sometimes protuberant, and from shape or armature, characteristic. This part, however, is not always distinctly apparent, although a line or suture usually separates it above, from the face. At its lower extremity the *labrum*, or upper lip, articulates, over which it is sometimes produced; and it extends at each lateral apex to the base of the insertion of the mandibles. The *genæ*, or cheeks, descend from the vertex laterally, behind the compound eyes, to the cavity of the head which contains the lingual apparatus, when folded in repose. These cheeks, at their lower extremity, sometimes embrace the articulation of the mandibles.

The *antennæ*, or feelers, are two filamentary organs articulating on each side of the face and above the clypeus. They comprise the *scape* (*a*), or basal joint, and (*b*) the *flagellum* or terminal apparatus; the latter consists of closely attached conterminal joints, and usually forms an elbow with the scape; collectively these joints number twelve in the female and thirteen in the male. They are all of various relative lengths, which sometimes aid specific determination. The scape, however, is usually much longer than any of the rest, and in some males has a very robust and even angulated shape. A description of the antennæ always enters into the generic character; they usually differ very materially both in length and form in the sexes. They are often

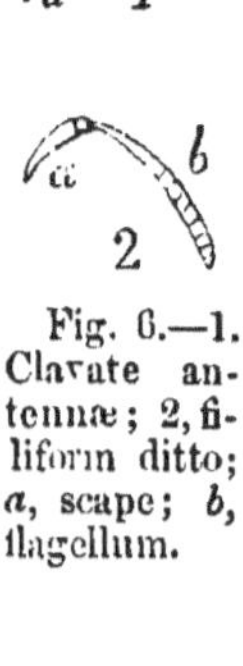

Fig. 6.—1. Clavate antennæ; 2, filiform ditto; *a*, scape; *b*, flagellum.

filiform (2), but more generally subclavate (1), and sometimes distinctly so, and where they have the latter structure it is found in both sexes. They constantly differ in the species of a long genus (*Andrena, Normada, Halictus*). In the male of the genus *Eucera*, they have a remarkable extension, being as long as the body, whereas folded back they are rarely so long, or not longer than the thorax in other males, speaking in reference only to our native kinds. In the females they are not often longer than the head. It is in the males of the genus *Halictus* that they take the greatest extension. In the male of the genus *Eucera*, we also find the remarkable peculiarity of the integument of some of the joints being distinctly of an hexagonal structure,—a peculiarity often observable in natural structures. In this case it may refer to the sensiferous function of the organ, and to which I shall have occasion to revert when I speak of the senses of our insects. We sometimes find the joints of the antennæ moniliform, something like a string of beads, or with each separate joint forming a curve, or with their terminal one, as in *Megachile*, greatly compressed.

The relative lengths of the joints often yield conclusive separative specific characters, and which may be very advantageously made available, especially where other distinctive differences are obscure, and in cases where the practised eye observes a distinction of habit, evidently specific, although it is difficult to seize tangible characteristics.

The *trophi* are the organs of the mouth of the bee collectively. When complete in all the parts, as exemplified in the genus *Anthoptera*, they consist of the *labrum*, or upper lip; the *epipharynx*, or valve, falling over and closing the aperture of the gullet; the *pharynx*, or gullet, which forms the true mouth and entrance to the œsophagus; the *hypopharynx* which lies immediately below the gullet and assists deglutition; the *labium*, or lower lip, and the true tongue. These parts are all single; the parts in pairs are the *mandibles*, the *maxillæ*, the *maxillary palpi*, the *labial palpi*, and the *paraglossæ*.

The *labrum*, or upper lip, is attached by joint to the apex of the *clypeus*; it has a vertical motion, and falls over the organs beneath it, in repose, when it is itself covered by the mandibles. It is usually transverse in form, but is sometimes perpendicular, especially in the artisan bees. It takes many forms, sometimes semilunar or linear, emarginate or entire, convex, concave, or flat, and is occasionally armed with one or two processes, like minute teeth projecting from its surface, but of what use these may be we do not know. In the female of *Halictus*, it has a slightly longitudinal appendage in the centre. It is usually horny, but is sometimes coriaceous or leathery. This labrum often yields good specific characters.

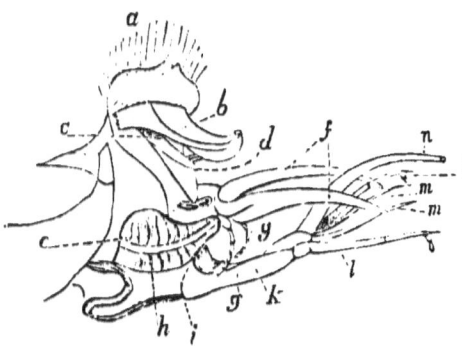

Fig. 7.—Trophi and their unfolding. *a*, labrum; *b*, epipharynx; *c*, pharynx; *d*, hypopharynx; *e*, mandible; *f*, maxillæ; *g*, maxillary palpi; *h*, mandible; *i*, cardium; *k*, labium; *l*, labial palpi; *m*, paraglossæ; *n*, tongue.

The *pharynx*, or gullet, is a cavity immediately beneath the *epipharynx*, which articulates directly under the base of the labrum, and which closes the pharynx from above, and immediately beneath this cavity is another small appendage, almost triangular, which receives the food or honey from the canal conveying it from the tongue, or directly from the mandibles, when it is masticated, and helps it forward to the pharynx to be swallowed. The *epipharynx* closes this orifice from above, the *labrum* then laps over it and the articulation of the lingual apparatus, both which are further protected in repose by the mandibles closing over the labrum. This triple protection shows the importance nature attaches to these organs. The more direct portions of the lingual apparatus are the *labium*, or lower lip, which forms the main stem of the rest, and articulates beneath the hypopharynx, and is beneath of a horny texture; it forms a knee or articulating bend at about half its length, and has a second flexure at its apex, where the true tongue is inserted. This *labium* is extensible and retractile at the will of the insect, and lies inserted within the under cavity of the head when in complete repose, and the insect can withdraw or extend a portion or the whole at its pleasure. Attached on each side, at its first bend or elbow, lie the *maxillæ*, which, for want of a better term, are called the lower jaws, and perhaps properly so from the function they perform; for at the point of their downward flexure, which occurs at the apex of the labium, and where the true tongue commences, they each extend forward in a broad, longitudinal membrane, partly coriaceous throughout its whole length, and these, folded together and beneath, form the under sheath of the whole of the rest of the lingual apparatus in repose, and often lap over its immediate base when even it is extended. Externally continuous, the line of these *maxillæ* is broken at the point of flexure at the apex of the *labium*, by a deep sinus or curve, and within this is inserted the first joint of the maxillary palpi. The portion of the maxillæ extending forwards, hence takes several forms, usually tapering to an acute point, but sometimes rounded or hastate, according to the structure of the tongue, to which they form a protection.

The *maxillary palpi* are small, longitudinal joints, never exceeding six in number, and generally in the normal or true bees not so numerous. They vary in relative length to the organ to which they are attached, and usually progressively decrease in length and size from the basal ones to the apical, but each joint, excepting the terminal one, is generally more robust at its apex than at its own special base. The function of these maxillary palpi is unknown. They are always present in full number in the *Andrenidæ*, and in some few genera of the true bees, but they vary from their normal number of six to five, four, three, two, and one in the latter; and it is curious that they are most deficient in those bees having the most complicated economy, as in the artisan bees and the cenobite bees; they thus evidently show that it is not a very paramount function that they perform. On each side, at the apical summit of the *labium*, are inserted the *labial palpi*. These are invariably four in number, but vary considerably in length and substance. In the *Andrenidæ* they have always the form of subclavate, robust joints, and are usually as long as the tongue, but not always; they are only half the length of that organ in the subsection of the acute-tongued *Andrenidæ*. In the normal bees, even in the genus *Panurgus*, which is the most closely allied to the *Andrenidæ*, the labial palpi immediately take excessive development, especially in their two basal joints, and the structure of these two joints, excepting in this genus and in *Nomada*, partakes of a flattened form and membranous substance. All these four joints are either conterminal, or the two apical ones, or one of them is articulated laterally, towards the apex of the preceding joint. These two are always very short joints, and are comparatively robust.

The *labial palpi* are, in the majority of cases, about half or two-thirds the length of the tongue, but in *Apathus* and *Apis* they are of its full length. At the immediate base of the tongue, and attached to it laterally, rather than to the apex of the *labium*, are the *paraglossæ*, or lingual appendages, which are membranous and acute, except in the *Andrenidæ*, where, in some, their apex is lacerated and fringed with short hairs. These organs are always present in the *Andrenidæ* and generally in the *Apidæ*, where they usually obtain extensive relative development; but in the artisan bees they are all but obsolete, and in *Ceratina*, *Cælioxys*, *Apathus*, and *Apis*, they are not even apparent. Their use also has hitherto eluded discovery, but that they are not essential to the honey-gathering instinct of the bee is especially proved by the latter instance.

The true tongue is attached to the centre of the apex of the labium, having the paraglossæ, when extant, and the labial palpi at its sides. In the *Andrenidæ* it is a flat short organ of varying form, either lobated, emarginate, acute, or lanceolate; but in the *Apidæ*, with *Panurgus* it immediately becomes very much elongated, and with this genus the apparatus whereby the tongue folds beneath obtains its immediate development; but this development exhibits itself most fully in the genus *Anthophora*. The tongue is usually linear, tapering slightly to its extremity, and terminating in some genera with a small knob. It is clothed throughout with a very delicate pubescence, which enables the bee to gather up the nectar it laps. That it should be called the lip seems an absurdity, for it exercises all the functions of a tongue, and

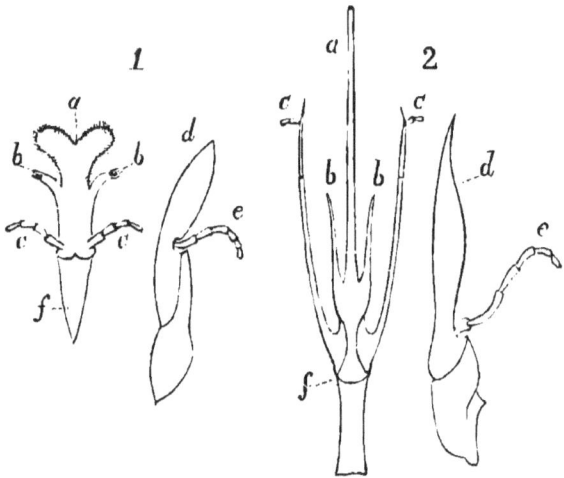

Fig. 8.—Extremes of structure of tongues: 1, in subnormal bees (*Colletes*); 2, in normal bees (*Anthophora*). *a*, tongue; *b*, paraglossæ; *c*, labial palpi; *d*, maxillæ; *e*, maxillary palpi; *f*, labium.

it would seem almost that the fine hairs, with which it is covered, are the papillæ of taste. Its structure in some genera seems to be a spiral thread twining closely round and round, but in others it appears throughout identical.

This tongue was formerly thought to be tubular, and that the bee sucked the honey through an aperture at its apex. The knowledge of the flat form of the tongues of other bees should have dissipated the illusion, for we could have been perfectly sure of the analogical structure and function of an organ in creatures so nearly alike. Réaumur's patient observations have totally dissipated the mistake, and through him we exactly know how the bee conveys the honey into its stomach'

As it exhibits an agreeable instance of the persevering industry and unblenching patience with which he made his researches, I will give a summary of what he says, for his bulky volumes, although teeming with delightful instruction, pleasantly narrated, will necessarily not be in every entomologist's hand, and where not, not even always readily accessible. His observations were made upon the honey-bee, but we may attribute the same mode of collecting to all the rest. He says:—When this tongue is not lapping the nectar of flowers but in a state of perfect repose it is flattened. It is then at least three times broader than thick, but its edges are rounded. It gradually narrows from its base to its extremity. It terminates in a slight inflation, almost cylindrical, at the end of which there is a little knob, which appears perforated in the centre. From the circumference of this knob tolerably long hairs radiate, and the upper side of the tongue is also entirely covered with hairs. The basal and widest portion above seems striated transversely with minute lines closely approaching each other.

The upper side of the anterior portion of the tongue seems of a cartilaginous substance, but the under side of the same part appears cartilaginous only over a portion of its width. The centre is throughout its whole course more transparent than the rest, and seems membranous and folded. It is only necessary to press the posterior portion of this trunk, whilst holding its anterior part closely to a light, towards which its upper surface must be turned, and then upon examining its inner surface with a lens of high power, a drop of liquid may be soon observed at its foremost portion. By continuing to press it this drop is urged forward, and as it passes every portion swells considerably, and the two edges separate more widely from each other. The under side of the tongue, which was before flat, rises and swells considerably, and all that thus rises up is evidently membranous. It looks like a long vessel of the most transparent material. But whilst this great increase of bulk is made upon the lower surface, the upper surface swells only a little, which seems to prove that its immediate envelope is not capable of much distension.

If a bee be observed whilst sipping any sweet liquor, the anterior portion of its trunk will be sometimes seen more swollen than when in action, and alternations will be observed in it of varying expansion.

The posterior portion of the trunk is a great deal larger than the anterior, and it is only in repose that the former nearly equals the latter in length. This posterior portion (this is the portion treated above as the *labium*, or under lip) is joined to the anterior by a very short ligature, wholly fleshy, and very flexible, which permits the folding of the trunk, and then its under side is quite scaly, very shiny, and rounded (the maxillæ). This portion is apparently more substantial than the rest. Its diameter gradually increases as it recedes from about the middle to about two-thirds of its length; there it is a little constricted, and the first of the two pieces of which it is composed there terminates. The first piece is rounded, for the purpose, it would appear, of fitting itself upon another, which serves as its base and pivot. This base is conical and of a scaly texture, and terminates in rather an acute point. It is this point which is articulated at the junction of the two small elongate portions of which we spoke at the commencement, and which carry the trunk forward.

In repose, the posterior part of the trunk lies along the lower part of the mouth, and the anterior part is folded back upon it, when it is covered by the maxillae, which then seem to form a portion of it. It has further another interior envelope; these are the two first joints of the labial palpi (in the *Apidæ*), which are entirely membranous, and these in repose cling closely to the tongue laterally.

The bee would certainly not collect its honey differently from a flower than it would from a glass wherein it might be placed to observe the process; and here it never appeared to obtain the honey by suction. The bee was never observed to place the end of its tongue in the drop of syrup, as it would necessarily do if it were requisite to imbibe it through what seems the small aperture at the extremity of the knob, at the end of the tongue, previously described. As soon as the bee finds itself near the spot spread with honey or syrup, it extends its tongue a line or so beyond the end of the palpi, which continue to envelope it throughout the rest of its length. If the honey be spread over the glass, the anterior portion of the tongue, which is exposed, is turned round that its superior surface may be applied to the glass. There this portion does precisely what the tongue of any animal would do in lapping a liquid. This tongue repeatedly rubs the glass, and, moving about with astonishing rapidity, it makes hundreds of different inflexions.

If the drop of syrup presented to the bee be thicker, or if it meet with a drop of honey, it then thrusts the anterior portion of its tongue into the liquid, but apparently only to use it as a dog might do its tongue in lapping milk or water. Even in the drop of honey the bee bends the

end of its tongue about, and lengthens and shortens it
successively, and, indeed, withdraws it from moment to
moment. We then observe it not merely lengthen and
shorten this end, but it is also seen to curve it about,
causing from time to time the superior surface to become
concave,—to give, as it were, to the liquid with which
it is loaded a downward inclination towards the head.
In fact, this portion of the trunk appears to act as a
tongue, and not as a pump. Indeed its extremity,
where the aperture for receiving the liquid is assumed
to be, is repeatedly above the surface of the liquid which
the insect is lapping.

By these continuous motions this anterior extremity
of the tongue charges itself with the nectareous fluid,
and conveys it to the mouth. It is along the upper
surface of this pilose tongue that the liquid passes.
The bee strives especially to load and cover it with
honey. In shortening the tongue to the extent, sometimes, of withdrawing it entirely beneath its sheaths, it
conveys and deposits the liquid with which it is charged
within a sort of channel, formed by the upper surface of
the tongue and the sheaths which fold over it. Thus,
these sheaths are, perhaps, less for the purpose of covering
the tongue than to form and cover the channel by which
the liquid is conveyed to the mouth. I have previously
remarked that the trunk can swell and contract; these
swellings and constrictions are observed to succeed each
other, and may be for the purpose of urging the liquid,
already in transit beneath the sheaths, forward towards
the true mouth. Further, I moved the sheaths aside
from their position above the tongue of a bee which I
held in my fingers, and I succeeded, by means of the
point of a pin, in placing an extremely small drop of

honey upon the tongue of this bee at a spot where it could be covered by the extremities of the external sheath. I then let these sheaths loose. Sometimes they spontaneously resumed their previous position, and sometimes I assisted them to resume it. The drop of honey which they then covered has in no instance returned to the extremity of the tongue; it has always passed towards the mouth, and doubtless entered that orifice itself. It is therefore very certain that the bee imbibes its honey by lapping, and that it never passes through the aperture which has been supposed to have been seen at the extreme apex of the tongue. Did this aperture really exist, it would be of extreme minuteness, and it did not appear to me possible that a large drop of honey, which I have seen imbibed in a very few instants, could in so short a time have passed by so minute an opening. A further confirmation of the non-existence of this orifice has been given me when, by pressing a tongue towards its origin to compel it to swell, I have detected the liquid which gave it its extension, but all my pressing would never make the liquid pass through the extremity, although the pressure has sometimes made it almost rend the membranes, to give it an opening to escape by. Having thus passed through the œsophagus into the stomach, it is then regurgitated into its requisite repository upon arriving at home.

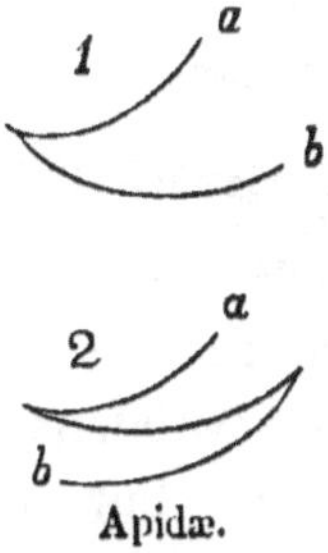

The entire proboscis, with all its appendages attached, has in the *Apidæ* three distinct hinges or articulations, including that which attaches it by its extreme base to

Fig. 9.—Mode of folding the tongue in repose. 1. In abnormal bee. 2. In normal bee. *a*, point of articulation beneath the hypopharynx: *b*, apex of the tongue.

the under surface of the mouth and lower portion of the head, the cavity of which, when folded, it fills, and even then the apex of the tongue protrudes in some genera beyond the sheathing maxillæ. In the *Andrenidæ* it has but two articulations, and the maxillæ always cover them entirely in repose. The first articulation, forming the fulcrum of the whole, is always elbowed in the *Apidæ*, and consequently not capable, like the rest of the joints, of full linear extension. The attached diagram will give a clearer conception of the mode of folding: *a* is the labium, and *b* the tongue.

As we have no complete description of the mode by which the tongue of the bee is worked, and how it gathers up its honey, I thought it desirable to be fuller upon the subject than was originally my intention.

The last portion of the *trophi*, also double, are the mandibles; they articulate on each side with the cheeks; they act laterally, and are variously formed, according to the economy of the insect. In the females they are usually more or less toothed,

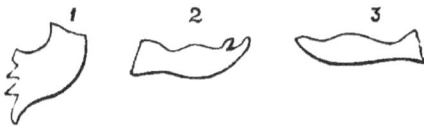

Fig. 10.—Mandibles: 1, of leaf-cutter bee (*Megachile*); 2, of burrower (*Andrena*); 3, of parasite (*Nomada*).

and are especially broad, curved, and toothed in the artisan bees. In *Apis* and *Bombus* they are subdentate. In males they are frequently simply acute, but in some species, especially in *Andrena*, they have a long spine at the base, which points downwards when they are closed. To this sex they appear to be of no use beyond aiding them to stay the wayward caprice or flight of their mistresses; and, although they have an analogical structure in the males of those genera wherein they are much dilated and toothed, yet they do not seem to be at all used by that sex for any purpose but sexual. In the females they are used for the construction of their burrows and nests, and for the purpose of nipping the narrow spurs and tubes of flowers to get at the nectar; and they often nip, whilst seeking pollen, the anthers of the flowers which have not yet burst their receptacles of pollen.

These insects must necessarily nicely appreciate the quantity of pollen requisite to the full development of the young insect, and, although we often observe a remarkable difference of size in the individuals of a species, this may rather arise from some defect in the quality of the nutritive purveyance than in its quantity, for instinct would as efficiently provide for this purpose as it unquestionably guides to the collection and storing of the nutritive supplies.

Having thus completed the description of the head and of all its attachments, I proceed to—

The THORAX, which is divided by sutures into three parts already mentioned above, viz. the *prothorax*, the *mesothorax*, and the *metathorax*.

The collar, or upper part of the prothorax, is often very distinct, and even angulated laterally in front, and frequently presents, both in colouring and form, a specific character. At its under portion on each side the anterior legs are articulated.

All the legs comprise the *coxa*, or hip-joint; the *trochanter*, which is a small joint forming the connection between this and the next joint the *femur*, or thigh; the *tibia*, or shank; and the *tarsus*, or foot. The latter consists of five joints, declining in length from the first, which is generally as long as all the rest united together; the first, in the anterior pair, being called the *palmæ*, or palms; and in the four posterior *plantæ*, or soles; the other joints are called the *digiti*, or fingers, or tarsus collectively; at the extremity of the terminal one are the two claws, which are sometimes simple hooks, but usually have a smaller hooklet within; they have both lateral and perpendicular motion, and between their insertion is affixed the *pulvillus*, or cushion. The *coxæ* in their occasional processes exhibit very useful specific characters, as do the markings and form of the remaining joints of the leg and foot, which in several genera furnish generic peculiarities. The four anterior tarsi have each a moveable spine, or spur, at their apex within, which can be expanded to the angle at which the insect wishes to place the limb, and to which it forms a collateral support; the posterior tibiæ have two each of these spurs, excepting in the genus *Apis*, which has none to this leg. Attached to this spur on the anterior tibiæ of all the bees, there is, within, a small *velum*, or sail, as it has been called; this is a small angular appendage affixed within the spur by its base. At the base of the palmæ of the same legs, and opposite the play of this velum, there is a deep *sinus*, or curved incision, the *strigilis*, called thus or the curry-comb, from the pecten, or comb of short stiff hair which fringes its edge. Upon this aperture the *velum* can act at the will of the insect, and combined they form a circular orifice. The object of this apparatus is to keep the antennæ clean, for the insect, when it wishes to cleanse one or the other of them, lays it within this sinus of the palma, and then, pressing the

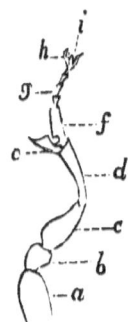

Fig. 11.—Anterior leg. *a*, coxa; *b*, trochanter; *c*, femur, or thigh; *d*, tibia, or shank; *e*, spur and velum; *f*, planta and strigilis; *g*, digitus; *h*, claw; *i*, pulvillus, or cushion.

velum of the spur upon it, removes, by the combined action of the comb and the velum, all excrescences or soilure from it, and this process it repeats until satisfied with the cleanliness of the organ: and this it may be frequently seen doing. This arrangement proves how essential to the well-being of the insect is the condition of its antennæ, the sinus, or strigilis, or curry-comb, as it may be called, being always adapted in size to the thickness of the antennæ, for insects being always both right- and left-handed, they therefore use the limb on each side to brush the antenna of that side. The palmæ and other joints of the tarsus of the fore legs are greatly dilated in many males, or fringed externally with stiff setæ, which give it as efficient a dilatation as if it were the expansion of its corneous substance. The anterior tarsi of the females are likewise fringed with hair, to enable them to sweep off and collect the pollen, and to assist also in the construction and furnishing of their burrows. The intermediate tarsi are as well often very much extended in the males, being considerably longer than those of the other legs. The use of the claws at the apex of the tarsi is evidently to enable the insect to cling to surfaces.

The manner in which the bee conveys either the pollen, or other material it purposes carrying home, to the posterior legs, or venter, which is to bear it, is very curious. The rapidity of the motions of its legs is then very great; so great, indeed, as to make it very difficult to follow them; but it seems first to collect its material gradually with its mandibles, from which the anterior tarsi gather it, and that on each side passes successively the grains of which it consists to the intermediate legs by multiplicated scrapings and twistings of the limbs; this then passes it on by similar manœuvres, and deposits it, according to the nature of the bee, upon the posterior *tibiæ* and *plantæ*, or upon the *venter*. The evidence of this process is speedily manifested by the posterior legs gradually exhibiting an increasing pellet of pollen. Thus, for this purpose, all the legs of the bees are more or less covered with hair. It is the mandibles which are chiefly used in their boring or excavating operations, applying their hands, or anterior tarsi, only to clear their way; but by the constructive or artisan bees they are used both in their building and mining operations, and are worked like trowels to collect moist clay, and to apply it to the masonry of their habitations.

The *mesothorax*, or central division of the thorax, has inserted on each side near the centre the four wings, the anterior pair articulating beneath the *squamulæ*, or wing scales, which cover their base like an epaulette, and this wing scale often yields a specific character. In repose the four wings lie, horizontally, along the body, over the abdomen, the superior above, the inferior beneath. The wings themselves are transparent membranes, intersected by threads darker than their own substance, called their nervures, which are supposed to be tubular. These nervures and the spaces they enclose, called cells, are used in the superior wing only, and only occasionally, as subsidiary generic characters, and their terminology it will be desirable to describe, as use will be made subsequently of it. At the same time, to facilitate the comprehension of the terms, an illustrative diagram is appended; but those parts only will be described which have positive generic application. I may, however, first observe that upon the expansion of the wings in flight, the insect has the voluntary power of making the inferior cling to the superior wing by a series of hooklets with which its anterior edge is furnished at about half the length of that wing, which gives to the thus consolidated combination of the two a greater force in beating the air to accelerate its progress. That the insect has a control over the operation of these hooklets is very evident, for, upon settling, it usually unlocks them, and the anterior are often seen separated and raised perpendicularly over the insect; but that this can be mechanically effected also is shown sometimes in pinning a bee for setting, when by a lucky accident the pin catches the muscles which act upon the wings, and they become distended, as in flight, closely linked together. Both the diagram and the description of this superior wing I borrow from an elaborate paper of my own in the first volume of the 'Transactions of the Entomological Society of London,' wherein I gave a tabulated view, in chronological order, of the nomenclature introduced by successive entomologists in the use they made of the anterior wing of the *Hymenoptera* for generic subdivision, and which I subsequently applied to my own work upon the 'Fossorial Hymenoptera of Great Britain.'

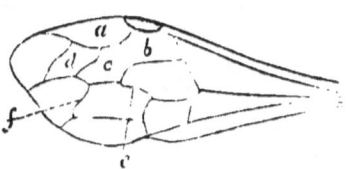

Fig. 12.—Superior wing. *a*, marginal cell; *b*, first cubital or submarginal cell; *c*, second ditto; *d*, third ditto; *e* and *f*, first and second recurrent nervures.

Attached to the mesothorax in the centre, above and behind, are the scutellum and post-scutellum, which in colouring or form often yield subsidiary generic or specific characters. On each side of the mesothorax in front, above the pectus, or breast, and just below and before the articulation of the anterior wings, there is a small tubercle, or boss, separated from the surrounding integument by a suture, the colouring of which frequently yields a specific character, but its uses are not known.

The *metathorax* carries the posterior legs laterally beneath, and in the centre, behind, the abdomen. The posterior legs are the chief organs used by the majority of bees for the conveyance of pollen to store in their cells, or, as in the case of humble-bees or the hive bee, the bee bread for the food for the young, or the requisite materials, in the majority of other bees, for nidification. To this end they are either densely clothed with hair throughout their whole extent,—usually externally only,—or this is limited to the external surface of the posterior shank. In the social bees this shank is edged externally with stiff bristles. In these, as in most of the bees, this limb greatly and gradually expands towards its articulation with the *planta*, or first joint of the tarsus; and this surface, which is perfectly smooth, serves to the social bee as a sort of basket to hold and convey the collected materials. The first joint of the tarsus, or planta, of this leg is also used in the domestic economy of the insect to assist in the same object. In the domestic bee the under side of the posterior plantæ have a very peculiar structure, consisting of a series of ten transverse broad parallel lines of minute dense but short brushes, which

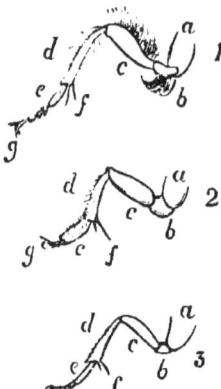

Fig. 13.—Posterior legs: 1, of abnormal bee (*Andrena*); 2, scopuliped normal bee (*Eucera*); 3, parasitic bee (*Nomada*). *a*, coxa; *b*, trochanter, with flocculus; *c*, femur; *d*, tibia; *e*, planta; *f*, spinulæ; *g*, tarsus, with its claws.

are used in the manipulations within the hive. Neither the queen bee nor the drone have this structure, and in the humble-bee and scopuliped bees the same joint is uniformly covered with this brush without its being separated into lines.

The ABDOMEN of bees has many shapes, its form being elliptical, cylindrical, subcylindrical, clavate, conical or subconical, and sometimes semicircular, or concavo-convex. It consists of six imbricated plates, called segments, in the female, and of seven in the male; in the latter sex, in several genera, it takes beneath at its base and at its apex, as well as at the extremity of the latter, remarkable forms and armature. It is very variously clothed and coloured, and sometimes extremely gaily and elegantly so; these various markings often giving the insects their specific characteristics; the clothing of the under side of this segment of the body, likewise, furnishes subsidiary generic characters, especially in the artisan bees, in whom it takes the place of the posterior legs as a polliniferous organ. This is possibly because were the supply conveyed *upon* their posterior legs it would be rubbed away as they entered the narrow apertures of their nests. Nature does nothing in vain, and there is evidently a purpose in this arrangement.

If we can trace peculiarities of structure to efficient reasons, differences of form may be rationally concluded as having their cause too, even if it elude our explanatory research. Although the reason of peculiar structure is not always obvious, it must exist, though undetected; as, for instance, why in some bees, as in *Megachile, Osmia, Chelostoma, Anthidium*, etc., the under side of the abdomen should be furnished densely with hairs to carry their provision of pollen home to their nest, when in other bees, as in *Dasypoda, Panurgus, Eucera, Anthophora*, etc. etc., it is conveyed upon the posterior legs, we do not know; we can only surmise that it is either to save the insect, in the former case, the labour of constructing a larger cylinder for nidification, so to prevent the possibility of its being rubbed off from the external surface of the legs, did these carry it, in entering the burrow, it being protected from this abrasion by being placed beneath the venter. In such insects the abdomen is usually truncated at its origin, or even hollowed within its base, thus to meet the projection of the metathorax, enabling it to draw itself closely up together, making the abdomen and metathorax, as it were, cohere. A different form of abdomen occurs in those bees which carry the pollen on their posterior legs. It is then more or less elliptical or lanceolate, which form permits the legs to be drawn up towards the metathorax within the space that kind of form furnishes, which, by this different but equivalent arrangement, meets the same object. The similarity of the adjustment of the abdomen to the metathorax to that of *Megachile*, etc. in *Apis* and *Bombus*, by which insects the provision is also carried on the posterior legs, results from the totally different economy and habitation of the social bees, to which this structure is necessary for many purposes.

If we observe this same peculiarity of structure in the cuckoo, or parasitical bees, it is because we find resemblances where there are alliances. Thus, the male artisan bees, although not assisting in the labour of constructing the apartments, have similarly dilated mandibles to those of their females. So also, in the form of the abdomen, the *Nomadæ* are like the *Andrenæ* and *Halicti*, upon which they are chiefly parasitical.

Melecta resembles *Anthophora* ; *Cœlioxys* has the form of *Megachile*, both in the hollowed base of the abdomen and the peculiar manner the latter has of raising its extremity,—something like a *Staphylinus*. Many other peculiarities of resemblancee might be enumerated.

Having thus completed the description of the external anatomy of the bee desirable to be known for facilitating the comprehension of what I may have subsequently to say. I shall now refer to a few peculiarities of their manners, which could not be conveniently introduced elsewhere.

In their modes of flight bees vary considerably; some dart along in a direct line, with almost the velocity of lightning, visit a flower for an instant, and then dart off again with the same fleetness and vivacity, like *Saropoda* and *Anthophora*; others leisurely visit every blossom, even upon a crowded plant, with patient assiduity, like *Bombus*; and some, either from fatigue, or heat, or intoxication, repose, like luxurious Sybarites, within the corolla of the flower. The males seem to flutter about in idle vagrancy, and may be often observed enjoying themselves upon some fragrant hedge-row. But the domestic bee and the humble-bee are the most sedulous in their avocation, and both cheering their labour with their seemingly self-satisfied and monotonous hum.

Bees, too, have a voice; but this voice does not proceed from their mouth, nor is it the result of air passed from the lungs through the larynx, and modulated by the tongue, teeth, and lips; for bees breathe through spiracles placed laterally along the several segments of the body, and their interior is aerified by tracheæ, which ramify variously through it; but their voice is produced by the vibration of the wings beating the air during flight. Even as Linnæus constructed a floral clock to indicate the succession of hours by the expansion of the blossoms of flowers, so might a Beethoven or a Mendelssohn—the latter in the spirit of his philosophical ancestor—note down the several sounds of the hum of the many kinds of bees to the construction of a scale of harmonic proportions, whose Æolian tones, heard in the fitfulness of accidental reverberation amidst the solitudes of nature, repeatedly awaken in the mind of the entomologist the soothing sensation of a soft, voluptuous, but melancholy languor, or exhilarate him with the pleasing feeling of brisk liveliness and impatient energy.

It is rarely that a bee is seen to walk, although a humble-bee or hive bee may be seen crawling sometimes from flower to flower on the same footstalk, but they are never good pedestrians. They convey themselves upon the wing from blossom to blossom, and even on proceeding home they alight close to the aperture of their excavated nidus, to which an unerring instinct seems to guide them. There occasionally they will meet with the intrusive parasite, to whom some genera (*Anthophora, Colletes*) give immediate battle, and usually succeed in repulsing the interloper, who patiently awaits a more favourable opportunity to effect her object.

Bees are exceedingly susceptible of atmospheric changes; even the passage of a heavy cloud over the sun will drive them home; and if an easterly wind prevail, however fine the weather may otherwise be, they have a sort of rheumatic abhorrence of its influences, and abide at home, of which I have had sometimes woful experience in long unfruitful journeys.

The cause would seem to be the deficiency of electricity in the air, for if the air be charged, and a westerly wind blow, or there be a still sultriness with even an occasionally overcast sky, they are actively on the alert, and extremely vivacious. They are made so possibly by the operation of the influence upon their own system conjunctively with the intensity of its action upon the vegetable kingdom, and the secretions of the flowers both odorous and nectarian.

Bees do not seem to be very early risers, the influence of the sun being their great prompter, and until that grows with the progress of the morning they are not numerously abroad. Early sometimes in the afternoon some species wend homewards, but during the greatest heat of the day they are most actively on the alert. The numbers of individuals that are on the wing at the same time must be astounding, for the inhabitants of a single colony, where they may, perhaps, be called semi-gregarious, from nidificating collectively within a circumscribed space, can be computed by myriads. And then the multitude of such colonies within even a limited area! When we add to this the many species with the same productiveness! Yet who, in walking abroad, sees them but the experienced entomologist? When we consider the important function they exercise in the economy of nature, and that but for them, in the majority of instances, flowers would expand their beautiful blossoms in abortive sterility, we can but wonder at the wise and exuberant provision which forecasts the necessity and provides accordingly. But that even these should not superabound, there is a counterbalance in the numerous *Enemies* to which they are exposed. The insectivorous animals, birds, among which there is one especially their arch-enemy—the bee-eater; those reptiles which can reach them; many insects in a variety of ways, as the cuckoo-bees, whose foster-young starve the legitimate offspring by consuming its sustenance; and personal parasites, whose abnormal and eccentric structure required an Order to be established for their admission. Strange creatures! more like microscopic repetitions of antediluvian enormities than anything within the visible creation, and to whose remarkable peculiarities I shall have occasion to return. Amongst the *Diptera* and *Lepidoptera* also they have their enemies.

Bees are sometimes exceedingly pleasant to capture, for many of them emit the most agreeable scents; some a pungent and refreshing fragrance of lemons; others the rich odour of the sweetest-scented rose; and some a powerful perfume of balsamic fragrance and vigorous intensity. These have their set-off in others which yield a most offensive smell, to which that of garlic is pleasant, and assafœtida a nosegay. These odours must have some purpose in their economy, but what it may be has not been ascertained.

They present very frequently remarkable disparities of structure and appearance in the sexes, so much so that its infrequency is rather the exception than the rule, and nothing in many cases but practical experience can associate together the legitimate sexes. Differences of size are the simplest conditions of these distinctions, for they occur also in individuals of the same sex. Differences of colour, consisting in increased intensity in the males, are also usually easily recognized; but the relative length and structure of the antennæ is a more marked disparity, and the development is always in favour of the male. The differences in the compound eyes are conspicuous in our native genera only in the drone, where they converge on the vertex, and throw the stemmata down upon the face. I have before alluded to special peculiarities in the legs when treating of those limbs. In the wings there are occasional differences, but so slight as not to require, in a general survey, special notice; but wherever they occur it is always in the male that the greatest extension of those limbs is found. The differences in the termination of the abdomen I have noticed above, and these sexual peculiarities in some genera are very marked. The spines which arm it in *Anthidium* and *Osmia*, and its peculiar structure in *Chelostoma* we can account for; but we have not the same clue to their uses in *Cœlioxys*, in which the action of the abdomen is upward, and not downward, as in the others.

The association of the legitimate partners of our native species has been to a great extent already accomplished and recorded; therefore, in this case, with the requisite guides to further instruction at hand, the commencing entomologist will find no obstruction, but may register the observations of his own experience to verify the discoveries of his predecessors.

It would seem from the facts that have been recorded, and the close investigations made, that in some instances the next year's bee is already disclosed and in the imago state, in the autumn of the existing year, so that it is ready, upon the first genial weather in the spring, to work its way out of its nidus, and take its part in the duties it has to perform. Whether this be for the economy of the food to the larva, or the saving of labour to the parent in gathering it, or that it would be prejudicial for it to lie dormant in the pupa state during the winter is not known, but thus in many instances it is.

Sometimes a late autumnal impregnation takes place, for the males of some *Andrenæ, Halicti,* and *Bombi* are found abroad only late in the autumn, and then in fine and recently disclosed condition.

It is a singular circumstance in the history of some species, that where they abound one season, nidificating on a certain spot in profusion, the following year, perhaps, and the year succeeding that, they will not be seen at all, but yet again a further year, and there they are as innumerable as ever.

What may control this intermittent appearance it is impossible to conceive, all the conditions of the spot and its surroundings being the same. This I have found to be a peculiarity incidental to many of the aculeate *Hymenoptera.* It occurs also in the flowering of many plants which blossom irregularly from season to season. It is a fact scarcely concordant with the observed rapidity of the disclosure of the larva from the egg, and the speedy growth, development, and transformation of the latter into the pupa and imago.

The wild bees appear to be of annual, or of even more restricted duration merely. Of this, however, we have no certainty. The conclusion is derived chiefly from the circumstance that, as they progressively come forth with the growth of the year, they, when first appearing, are in fine and unsoiled condition. There are evidently in some species two broods in the year; the one in the spring and the other autumnal. In bees without pubescence we have not the same guide. But humble-bees are reputed to have a longer life than of one year, and hive bees are said to survive several years, a duration of existence inconsistent with analogy, and which has been repeatedly and strongly denied.

In speaking of the *antennæ* and *palpi*, I have called them sensiferous organs. The organ necessarily implies the perception, or whatever it may be, conveyed to the sensorium through its means, this being the receptacle of the sensation or idea, the external organ communicates. It is thus that activity is given to a power of discrimination, and consequently of election or rejection by the creature. This sensorium, in the higher animals, is the brain; and in the lower, where the nervous system is very differently constituted, a ganglion, or knot of nervous substance. That this brain, or ganglion, is the power exercising the control, may not be admitted, although it is there that our research compulsively terminates. The power itself is essentially spiritual, acting through a material agent, and may be an efflux of this nervous mass. Whether it cease with the death of the organ, we have no means of knowing. That it may be in some way analogous in nature to the human mind, but to a limited extent, there is reason to surmise. This power, in its collective capacity, is called INSTINCT. This instinct is a faculty whose clear comprehension and lucid definition seem impossible to our understanding. Its attributes are very various, and its operations are always all but perfect. It is an almost unerring guide to the creature exercising it, and is as fully developed on its awakening as is, and with it, the imago upon its transformation.

Although observation has thought to have detected that experience sometimes uses a selection of means, and thus occasionally modifies the rigid exercise of the faculty, by adapting itself to the force of circumstances, it, when so, evidently assumes a higher character than has been willingly accorded to it. This instinct teaches the just disclosed bee, without other teaching than that of the intuitive faculty, where to find its food, and how to build its abode. It directs it to the satisfying its material needs, and instructs it to provide for its offspring, and to protect them whilst in their nidus; the impulse to which follows immediately upon the satisfaction of the sexual desire, to which it is the seal.

If it be *memory* that guides the bee from its wide wanderings back to its home, this then becomes an attribute to the faculty. Instinct indicates to them their enemies, and the wrongs these may intend, and shows them how they may be repulsed or evaded. In some of its operations it seems to be of a more perfect capacity than the operative faculty of human intelligence.

The senses evidently possessed by our insects are sight, feeling, taste, and smell, but whether they hear we cannot know, although the antennæ have been supposed to be its organ, for the apparent responsiveness of these to loud and sudden sounds, may equally result from the agitations of the air these produce. Their possession of touch, taste, and smell, are implied from what has been observed.

They certainly exercise a will, evinced by their power of discrimination, which decides what is salutary and what is noxious; and the passions are exemplified in their revenge, their sexual love, and their affection for their offspring, the latter being exhibited in their unremitting labour and careful provision for them, although they are never to see them. If there be any precedence in the order of the relative quality and distinction of the bees, it will be shown in the degree of superiority with which this function is accomplished. The perfection of this function we see progressively maturing as it passes onwards from the merely burrowing-bee to the more complicated processes of the masons, carpenters, and upholsterers,—all solitary insects, and working each individually and separately to the accomplishment of its object. But we may certainly inquire where we shall intercalate the sagacity of the cuckoo-bees. A vast bound is immediately made from the artisan bees to the social bees with three sexes, which, as first shown in the humble-bee, works in small and rude communities, with dwellings of irregular construction. The next and most perfect grade is the metropolitan polity, accomplished architecture, laborious parsimony, indomitable perseverance, and well-organized subordination of the involuntary friend of man, the domestic bee. This insect has furnished Scriptural figures of exquisite sweetness, poetry with pleasing metaphors, morality with aphorisms, and the most elegant of the Latin poets with the subject of the supremest of his perfect Georgics.

That bees feel pain may be assumed from the evidence we have of their feeling pleasure, although instances are on record of insects surviving for months impaled; and they lose a limb, or even an antenna, without evincing much suffering, and I have seen a humble-bee crawling along on the ground with its abdomen entirely torn away.

In speaking of the antennæ above, as possibly the organs of hearing, I would wish to add, that they evidently possess some complex function, of which, not possessing any analogy, we cannot certainly conceive any notion. They are observed to be used as instruments of touch, and that too of the nicest discrimination. They seem to be extremely sensitive to the vibrations of sound and the undulations of air, and keenly appreciative of atmospheric influences, of heat, of cold, and of electrical agitations. That they are important media in sexual communication must be assumed from their great differences of structure and size in the sexes, probably both as organs of scent and stimulation. I have often observed bees thrust their antennæ into flowers, one at the time, before they have entered the flower themselves, and in some insects, as in the Ichneumons, they are constantly in a state of vibration,—a tribe which, although of the same order, are remote in position from the bees, yet they may be instructively referred to by way of analogy in the discussion of the uses of an organ, whose functions so clearly follow its structure and position in the organization of the entire class of insects, that the analogy might be safely assumed in application to every family of the class, if observation could only correctly ascertain its uses in any one of them.

That it is of primary signification to the bees, is sufficiently shown by nature having furnished these insects with an apparatus designed solely to keep the antennæ clean, and which I have described above, when speaking of the structure of the anterior leg.

In the social tribes the antennæ are used as means of communication. The social ants, bees, and wasps may be often seen striking each other's antennæ, and then they will each be observed to go off in directions different from that which they were pursuing. An extraordinary instance of this mode of communication once came under my own notice, having been called to observe it. There was a dead cricket in my kitchen, another issued from its hole, and in its ramblings came across this dead one; after walking round, and examining it with its antennæ and fore legs a short time, it started off. Shortly, either attracted by sound, or meeting it by accident, it came across a fellow; they plied their antennæ together, and the result was that both returned to their dead companion, and dragged him away to their burrowing-place, —an extraordinary instance of intercommunication which I can vouch for.

It would be curious to know if the means of communication thus evidently possessed by animals, extends beyond the social and gregarious tribes, and whether the faculty undergoes any change through differences of climate and locality, as man has done in the lapse of time. For man, notwithstanding the vastly divergent differences of race, may be obscurely tracked through the dim trail of the affiliation of languages to one common origin. But the complete identity of habit throughout the world of those genera which are native with us, would seem to affirm that they are as closely allied in every other particular, were we in a condition to make the investigation, and whence we may conclusively assume that they all had one central commencement.

That this mode of communication, and this exercise of the organ in the solitary tribes is limited to the season of their amours is very probable, and I apprehend that it is not exercised between individuals of distinct species. But that, at that period, their action is intensified may be presumed from the then greater activity of the males, who seem to have been called into existence only to fulfil that great object of nature, and which she associates invariably with gratification and pleasure. Even in plants it may be observed to be attended with something very analogous to animal enjoyment in the peculiar development at that period of an excessively energetic propulsion, which is the nearest approach the vegetable kingdom makes to the higher phase of sensiferous life.

The clothing and colouring of bees are very various, but the gayest are the parasites, red and yellow, with their various tints, and white and cream-colour decorate them. The ordinary colour is deep brown, or chestnut, or black. Where the pubescence is not dense, they are often deeply punctured, and exhibit many metallic tinges. Many are thickly clothed with long hair, and this, especially in the *Bombi* and *Apathi*, is sometimes of bright gay colour, yellow, red, white, of a rich brown, or an intense black, sometimes in bands of different tints upon the same insect, and sometimes of one uniform hue.

CHAPTER III.

SKETCH OF THE GEOGRAPHY OF THE GENERA OF
BRITISH BEES.

In giving a broad sketch of the geography of the genera of bees which are native to our islands, but whose local distribution I shall reserve for notice in the account of the genera themselves, I must regret at the outset the lack of materials for its satisfactory treatment.

There are but very few exceptions to the dearth of assiduity in this direction; a very favourable one is that of the son of the late venerable hymenopterologist, the Count le Pelletier de St. Fargeau, who, at his military post as an officer of the French army in Algeria, stationed at Oran, collected energetically for his father in that district, and where, in one of his collecting excursions, he was severely wounded by a musket-ball. Another equally favourable exception is that of Sydney Smith Saunders, Esq., residing at Prevesa, in Albania, who has strenuously and perseveringly collected in that country. Here and there we can point to something having been done in Upper India, in the vicinity of Poonah, at Pondicherry, in Java, in some limited localities of China, and to some extent in Australia, Tasmania, and New Zealand, but nothing of any magnitude. There is much hope that a great deal has been done in Ceylon by Mr. Thwaites, who, when resident at Bristol, was a most ardent and successful hymenopterologist.

The Egyptian *Hymenoptera* have been extensively and admirably figured by Savigny, in the Imperial superb work published under the auspices of Napoleon I., but to these, unfortunately, no descriptive text was published, and they are therefore as useless to science as if they had not been figured. But those collected by Ehrenberg, and figured by Klug, in the 'Symbolæ Physicæ,' exhibit how rich in variety is that remarkable region. These figures may be called the *ne plus ultra* of entomological artistic skill.

Unfortunately, this Order has been sadly neglected for the sake of the less troublesome *Coleoptera*, and the more conspicuous *Lepidoptera*. This is plainly perceptible from the paucity of species recorded as having been once in the Count Dejean's collection, where we might have expected to have obtained a rich view of the *Hymenoptera* of Spain; as also in those of other French collectors, who have had rare but neglected opportunities for the purpose. It is true M. Brullé has done a good deal in Greece. We are, as yet, in comparative ignorance, from the same cause of neglect, of the *Hymenoptera* of Italy, excepting something that has been done by the Marquis Spinola, in Liguria, and by Rossi, in Tuscany. A little has been contributed towards that of Carniola, but we are almost ignorant of the *Hymenoptera* of Sicily, which, from various causes, are likely to be very peculiar. Mr. Swainson's collection of them, although not numerous, were neglected until they became unintelligible. The only European countries that have been tolerably gleaned are Germany, Sweden, a part of Russia, and even Finland. It is impossible for any entomologist to examine every locality for himself, he must, in great measure, depend on the labours of others; and, of course, I can only speak of the collections which are accessible to me, or which are described in monographs, or have been named in lists that have been published. Doubtless the Museum of Berlin, so long under the administration of a lover of the Order, Dr. Klug, would present a large contribution to our knowledge of the distribution of the forms, did a list of its riches exist. Such a list of the *menoptera* of Portugal, contained in Count Hoffmansegg's collection, was published many years ago in Illiger's 'Magazin der Insectenkunde..'

It has been a fatality incidental to this entomological branch of the study of natural history that some of its most energetic cultivators have been taken early away. There was formerly Illiger, then our own Leach, and then Erichsen. Leach, but for his afflicting malady, would have done much for the science; still, let us hope that the *Hymenoptera*, and especially the bees, are gaining ground in the estimation of entomologists generally, and that not many years will pass before collectors will possess them in abundance. For the present, I can but give a slight summary of the knowledge we possess on this subject.

Thus science has sustained great loss by reason of the unfortunate neglect which the family of bees, and, indeed, the Order of *Hymenoptera* generally, has met with from collectors in distant localities whose tastes have led so directly to the collection of other more favoured Orders, and the opportunities for repairing the consequences of such neglect being in some cases extremely rare. The present slight attempt to trace the geography and cosmopolitan range of our native genera of bees will necessarily be affected to some considerable extent by this neglect.

Although the materials in our possession will yield some fruit, yet their collection will be but the gleaner's handful, instead of a loaded wain from a rich and abundant harvest. As what I have gathered may still have an interest for some of my readers, I will lay it before them, and in doing so I shall take the genera in their methodical series.

The genus Colletes comes first, a position the more remarkable from the peculiarities of its economy and form, which bring it closely to the true bees, as do also its aptitude, by reason of its structure, for collecting pollen, and its energy in gathering it. The divergence in the form of the tongue brings it, however, to the extreme commencement of the series, it being the closest structural link we find for connecting the bees with the preceding family of wasps. This genus, in our own species, ranges through northern Europe to the high latitude of Finland, passing through Sweden; and it occurs also in Russia and in the Polish Ukraine. In other species than ours, and differing among themselves, it occurs at both extremities of Africa, in Egypt, and Algeria, and at the Cape of Good Hope; but whether throughout the wide interval collections do not inform us. It has been sent from Turkey, but whence?—for this is as vague a designation as Russia, both being empires which spread over vast areas,—and, if found in their Asiatic divisions, are the only instances we know of its Asiatic occurrence. It is so easy for collectors to add to their specimens a defined and precise locality, that its omission in any instance is to be regretted, as in many ways, and in all kinds of collections, it might be very serviceable to science. To our present purpose it has but a collateral interest as an object of curiosity, yet curiosity has led to many discoveries which have proved valuable to mankind. All the divisions of natural science have a mutual and convertible bearing, and closely interlink in their relations. Thus, insects denote the botany, which further indicates the climate or elevation and soil; and the superficial soil will point geological conclusions to subsoil and substructure. One natural science well mastered gives a key to the great storehouse of nature's riches, and yields a harvest of many different crops. This episode may be excused for the hint it is intended to give of the paramount importance of the correct registration of special localities.

The genus *Colletes* also occurs in the Canary Islands, which shows a trending tendency to its southern habitat at the Cape of Good Hope. It occurs on the western edge of South America, in Chili; it is found on its northern boundary in Columbia, and has been discovered in the southern States of North America, in Florida and Georgia; but there is no record of its further northern occurrence upon that continent. About thirty species are known.

The genus Prosopis, or as it is more familiarly known by the name of Hylæus, is found in some of our native species throughout France and Germany, and, like the preceding, as high up as Finland, through Denmark and Sweden, to the adjacent parts of Russia. It is remarkable that it is caught in Algeria, although not recorded as occurring in several of the southern European States. But the apparent restriction of some of our species to our own islands possibly arises from the fact of special attention having been paid to them in this country only.

The genus itself, in other and more variegated forms than ours, presents itself in some portions of southern and south-western Europe, where the highly ornamented species would point almost to the certainty of its being a parasitical genus, great decoration being in our native genera of bees the badge of parasitism, and may be indicative of those habits, combined as they are conjunctively with their destitution of polliniferous organs. Some of our native entomologists have, however, assumed, upon what appears to me very inconclusive grounds, that the genus is not parasitical. The observations, however, of the most distinguished French hymenopterologists confirm the notion of their being parasites, which appears strengthened by the argument above suggested with regard to colour.

This genus is apparently fond of hot climates. In eastern Europe, it occurs in Albania and the Morea, its extreme western domicile is Portugal, and its southern European habitat is Sicily. It is found in Algeria and Egypt, and at the Cape of Good Hope. We discover it in India, in the southern tropics at the Brazils, and in the northern tropics at the Sandwich Islands; and it ranges along the southern edge of Australia, from Swan River through Adelaide and Port Phillip to Tasmania. The United States of North America furnish it, and on that continent it seems to contradict its ordinary tropical inclination by being exceptionally found upon the confines of the arctic circle at Hudson's Bay. Nearly sixty well-distinguished species are recorded.

The genus SPHECODES has also a wide distribution.

Our native species are found throughout France and Germany, Greece and Spain, still one or two seem limited to our islands. The genus is recorded as in Albania, Algeria, and Egypt; it is found on the western edge of Africa at the Canaries; it occurs also in northern India, in the United States, on the western side of South America at Chili, and then we have a wide gap, for its next appearance is at Sydney, New South Wales. About twenty species are known.

The genus ANDRENA, although infinitely more numerous in species than the genus *Halictus*, which is also abundant, does not appear to have so wide a distribution as the latter. Peculiarities of habits possibly limit its diffusion, although nothing has occurred to naturalists to explain the circumstance, unless it be the adventitious fact of no specimens having fallen into the hands of the collector. Our own species, represented by one or several members, are found (although some seem restricted to England) throughout Europe, north and south, east and west, as also in its islands. In Africa it is seen in Algeria and Egypt, and it occurs in the Canaries; and in Asia it is found in Siberia, and in northern India; but we have no connecting chain to link those Asiatic and African localities,—although we may well suppose that it might be discovered amongst the steppes of Thibet and Tartary, revelling amidst the flowers of their luxuriant pastures, and even amongst the Persian sands. It passes through the United States from Florida up and to our own colony of Nova Scotia, and extends its range to Hudson's Bay. We do not trace it further. Nearly two hundred species occur.

The genus CILISSA, too, has a limited distribution, and occurs in the same countries, but ranges as high as Lapland; it also crosses the Atlantic, being found in the United States. About six are known.

Our solitary species of the genus MACROPIS, which is isolated possibly only from having been overlooked, appears to have but a European existence, and is found in France, Germany, Denmark, Sweden, and Finland.

The genus HALICTUS is very cosmopolitan. Some of our own species occur throughout Europe, excepting only Italy and Sicily, although they are to be found in Portugal and Dalmatia, thus traversing its entire breadth; but from the latter country they do not seem to range down to Albania and Greece, yet are they discovered in Malta, and even in southern Africa, but they have not been recorded as extant in northern portions of that continent. Other species have been sent from the western coast of Africa and the adjacent Canaries, with their adjunct, Madeira, and the genus ranges from Barbary through Senegal and Sierra Leone; some species also are found at the Cape of Good Hope.

On the other side of Africa the genus has been discovered at the Isle of Bourbon; it then takes a wide sweep, occurring first in northern India; it then springs up at Foo-chow-foo, and it is found in northern China. In western Asia it occurs in Syria. Across the Pacific it is found in Chili. Its next appearance on the rich and diversified continent of America is across its southern bulk, presenting itself in the Brazils, and on its northern boundary at Cayenne, and in Columbia; and it then appears again in Jamaica. In North America it occurs throughout the United States from Florida upwards, where the genus in its species has a very English aspect, and if they be dissimilar, as may be fairly surmised, they are so very like our own that one is said to be absolutely identical throughout Europe and in Ohio. It passes still forward and occurs in Nova Scotia, Hudson's Bay, and elsewhere in arctic America, where the botanist might almost herbalize through the agency of our insects, for the pollen they carry and still retain in cabinets would often indicate the plants which they there frequent. Thus those stern regions are not barren in fragrant and attractive beauties. We find it, too, in common with *Sphecodes* at Sydney, New South Wales, whence, doubtless, it passed to New Zealand, where it has been collected. About one hundred and fifty are registered.

With the next genus, DASYPODA, I terminate the geography of the *Andrenidæ*. Our own single species of these very elegant bees occurs throughout France and Germany, and abounds in Sweden. Other species, all elegant, occur in the Isles of Greece, in Albania, and the Morea; profusely at Malaga in Spain, and at the further extremity of northern Africa in Tunis, and in Egypt. Twenty are known.

The genus PANURGUS is the advanced guard of the true bees, for, although it still retains much of the appearance and structure of the terminal genus of the preceding sub-family of *Andrenidæ*, it is strictly distinct, and well links the two sub-families together. This very peculiar form is limited in number of species and in distribution, for five only have been recorded.

Our own species occur throughout France, Italy, Germany, Switzerland, Denmark, Sweden, and Finland, and one of them has also been sent from Oran. The genus is small, and may have been overlooked in other countries, although its appearance is sufficiently distinct and marked to have caught the eye. It is as lithe and active as a Malay, as black as a negro, and as hairy as a gorilla, looking like a little ursine sweep.

The genus EUCERA, of which we have but one representative, although considerably more than fifty species are known, has not so wide a range as might be expected from their numbers. Our own is found throughout Europe and in Algeria. Other species occur in Russia, the Morea, Albania, Dalmatia, and Egypt. In Asia some are found in Syria, and at Bagdad; and from the New World they have been sent from Cayenne and the United States.

The genus ANTHOPHORA, to which the genus *Saropoda* is very closely allied,—so closely, indeed, that by the celebrated hymenopterologist Le Pelletier de St. Fargeau the species of both are incorporated together,— has, even as now restricted, a world-wide dissemination, and numbers nearly a hundred and fifty species. Several of our own occur throughout France and Italy and the whole of northern Europe, and even among the Esquimaux in the arctic regions, showing that a bridal bouquet may be gathered even there; for where bees are flowers must abound.

The genus in other species shows itself in the south of Europe, viz. in Spain, Sicily, the Morea, and Dalmatia; by way of Syria and Arabia Felix it passes down to Egypt and occurs in Nubia and also in Algeria. It dots the western coast of Africa at Senegal and Guinea, and has been discovered in the Canaries, and again makes its appearance at the Cape of Good Hope, rounding it to Natal. It travels round the peninsula of India, being found at Bombay, in Bengal, and in the island of Ceylon, and passes onward by way of Hongkong to northern China, where, dipping to the Philippines, it next occurs in Australia. In the New World it is found on its western side at Chili, and traverses that continent to Paraguay and Pará, and has been sent from the West India Islands of Cuba, St. Domingo, and Guadaloupe. From Mexico, where we next find it, it passes to Indiana, and occurs throughout the United States, and thus completes its progress round the world. About one hundred and thirty are known.

The genus SAROPODA is closely allied to *Anthophora*, as closely as *Heriades* is to *Chelostoma*, and is very limited in numbers, ten only being known, and but one of which is native with us. The genus occurs throughout France and Germany, and has been sent from Russia, Egypt, South Africa, and Australia, thus having a very wide range notwithstanding the paucity of its species.

The very pretty genus CERATINA, although numbering but few species,—fewer than thirty,—and although not found in Australasia, is widely scattered throughout the Old and the New Worlds. Our own species inhabits as far north as Russia. Other species occur throughout France, and in the south of Europe, and show themselves in the Morea, and in Albania. North, South, and Western Africa possess the genus, it being found in Algeria and at the Cape of Good Hope, and in the intervening district of Senegal. It has been brought from Ceylon and Bengal, and also from the north of India. It reaches China by way of Java and Hongkong: and in the New World has been found in the Brazils and Cayenne, in the Southern, and throughout the United States in the Northern continent.

The genus NOMADA is the first of the genuine parasitical bees, and about the habits of which no doubt can be entertained; certainly not the same as attaches both to *Hylæus* and *Sphecodes*, among the *Andrenidæ*. The parasitical habits of *Nomada* are evident and unmistakable. This is the handsomest genus, in variety of colour and elegance of form, of all our native bees, but the species are never conspicuous for size. They have much of the appearance of wasps, and are often mistaken for them even by entomologists, who have not paid attention to bees. Many of our native species seem limited to our own islands: others of our species occur in France and Germany, and through Denmark in direct line to Lapland, turning down into Russia, and have been caught as far south as Albania. One of our species, or so like as to want distinguishing characteristics, is found in Canada. Did ours migrate there? and how? The genus is of wide distribution, but occurs only north of the Equator, where it spreads from Portugal to the Philippine Islands. It is found in Siberia and Northern China, whence through the Philippines it passes to Tranquebar, then up to Northern India, and thence by Bagdad to the Morea and Albania, and dips down to Northern Africa at Tunis, and on to Oran and Tangiers, and completes its circuit in Portugal. It is doubtless parasitical upon many more genera and species than we find it infest in this country, although all that the several species pair off with here are not fully designated, especially among the *Andrenæ*, and smaller *Halicti*. The number of species, British and foreign, known to collectors approximate to a hundred.

The genus MELECTA is another handsome parasitical insect. This is always a dark beauty, and is very limited in species, for, as far as they may be estimated from the contents of collections, its numbers do not reach twenty. Our own species occur throughout the whole of Europe, north and south. Others are found in Sicily, Albania, the Morea, and show themselves at Bagdad. The genus has been sent from the Canaries, and crosses the tropics into Chili, but does not seem to have occurred elsewhere in either North or South America, although one of the genera (*Eucera*) on which, with us, it is parasitical, is found in the latter country, and the other genus (*Anthophora*), which it also infests, is found throughout the world, excepting in Australasia. In all those countries, the closely-allied exotic genus *Crocisa*, which is very numerous in species, may supply its place.

The elegant genus EPEOLUS occurs in our own species throughout northern Europe, as high as Lapland, and is found also at the southern extremity of the continent of the Old World, at the Cape of Good Hope. It has been brought from Sicily, and other species come from Siberia. The genus in America passes down from the United States, by way of Mexico, to the Brazils, where it crosses the southern continent, having been transmitted from Chili. It is very limited in the number of its species, considering its wide diffusion, for not more than twenty are registered. It is almost identical in distribution with the genus *Colletes*, upon which it is with us parasitical. The species are never so large as those of the preceding genus, *Melecta*.

The genus STELIS is limited both in number of species and distribution, although the spots whence it has come are wide apart. Our own species are found throughout France and northern Europe, as far as Finland. Other species occur in North America, and the Brazils, but the whole number yet described is under ten.

The remarkable form in both sexes of the genus Cælioxys occurs in identity with our own species throughout France and Austria, and spreads north to Finland and Russia, and through all the intervening countries. It is singular that it should not be recorded from southern or south-western Europe, as it is found in Oran. Other species of the genus have been found in northern Africa, Egypt, and Algeria. On the western coast of Africa it has been caught on the Gambia, at Sierra Leone, and on the coast of Guinea. It doubles the Cape of Good Hope, where it is found extending its range to Port Natal. From Asia we have it from Turkey, and again from India. It has been sent from the hither side of South America, from the Brazils, and separately from Pará, and occurs at Cayenne, and in the West India Islands, Cuba, and St. Thomas's, and extends as high in North America, through the United States, as Canada. It is quite probable that it has as wide a range as the bees upon which it is parasitical (*Megachile*), although it has not yet come from such extensively-spread localities. More than fifty species are known, but some of our own have not yet been enumerated amongst those found elsewhere.

The genus MEGACHILE, which embraces the most renowned of the mechanical bees, is extremely cosmopolitan, spreading north and south, east and west; and is also very abundant in the numbers of its species, the census extending to not far short of two hundred. Some one, or several of our species, although other species are limited to our own country,—spread through Italy and France, and all the countries of northern Europe to the high latitude of Lapland, which is higher than where even one of ours (viz. the *M. centuncularis*) is again found, which occurs in Canada and at Hudson's Bay. The genus also frequents southern Europe, in Spain, Sicily, and Albania, and in the East, in the Caucasus and Dalmatia. It traverses Turkey by Bagdad to India, having been captured in Nepaul, and it descends southward in the Indian peninsula, where it has been found at Bombay. From India it stretches to the Mauritius, thence across the Indian Ocean to Java, and thence to Hongkong and northern China. It then dips to the Philippines, and doubtless through the islands of the Indian Archipelago to Australasia, from which continent none are registered from its northern and eastern settlements, but species abound along its southern edge from Western Australia, through Adelaide to Tasmania. The genus has been brought from the West India Islands, St. Thomas's, St. Croix, and Cuba : it is found upon the main from Mexico, descending to the Brazils. It skirts all the coasts of Africa, being discovered in Egypt and Algeria, along the western coast by the Gambia, Senegal and Sierra Leone to Guinea, and the island of Fernando Po, and then again occurs at the Cape of Good Hope. Ascending the eastern coast by Natal, it stretches to Abyssinia. The species are very abundant in India, Africa, and Australasia.

The genus ANTHIDIUM, although very numerous in species, and differing more remarkably in form amongst themselves than most other genera, has a far less extensive range, no species having been found in Australasia or India, although it occurs in Arabia, Syria, and Mesopotamia. Our own solitary species occurs in France, Italy, and the whole of northern Europe, extending to

Finland. In southern Europe the genus inhabits Sicily, Spain, the Morea, Albania, and Dalmatia, and is also very abundant in Southern Russia. In Africa it is found in Nubia and Algeria, and on its north-western edge in Barbary, whence it descends by the Gambia and Sierra Leone to the Cape of Good Hope, and thence reaches to Natal. It is then found in Chili, and crossing the South American continent occurs in the Brazils, whence it ascends to Cayenne, and, by way of Mexico, to the United States. The number of species recorded exceed a hundred.

The remarkable genus CHELOSTOMA is very limited in the numbers of its species, of which less than a dozen are known; as also in the extent of their distribution. Our own are found throughout northern Europe, as far as Lapland, and in Russia. In southern Europe they occur in the Morea, and the genus has been discovered in Georgia in North America.

The closely-allied genus HERIADES seems limited to a European habitation, and occurs only in our own solitary species, but it ranges, like the preceding, to the high latitudes of Lapland.

ANTHOCOPA seems limited to our own country and France, possibly only from its having been associated from similarity of general habit with the genus *Osmia*. Only one species appears to be known, but this has a world-wide celebrity, from the interesting account given by Réaumur, of its hanging its abode with symmetrical cuttings from the petals of the poppy.

The genus OSMIA, although not including such able artisans as *Megachile*, still has in its species very constructive propensities. Indeed, all the bees which convey the pollen on the under side of the abdomen, are more or less builders or upholsterers. The genus has a wide range, and is tolerably numerous, numbering more than fifty species. Some of our own occur throughout Europe, and, like the two preceding genera, are found in the highest continental latitudes. Some of ours also occur in Algeria and the Canaries, other species in Albania and Moravia. In Africa they are found in Egypt, Barbary, and Port Natal, and in the New World from Florida, in the United States, through Nova Scotia to Hudson's Bay.

The genus APATHUS, which is parasitical upon *Bombus*, and to the uninitiated has all the appearance of this genus, seems to be the only instance of a parasitical genus of bees so closely resembling the σῖτος, (as we may, perhaps, for the sake of avoiding a periphrasis, be allowed to call the bee upon which the parasite is found,) as to be so easily liable to be mistaken for it, and which was indeed the case by even such a sagacious entomologist as the distinguished Latreille; but Kirby had already noticed the difference, suggesting its separation from *Bombus*, until about the time that St. Fargeau was induced to propose a distribution of the Hymenoptera, based generally upon economy and habits, to which he had been led by a refining investigation of structure, that the distinguishing difference was appreciated, and used generically, by Mr. Newman. This difference, like many other simple facts, now that it has been found, is very obvious. It consists in the genus having no neuters, and the female of the species no polliniferous organs, but the determination of the legitimate males, by means other than empirical, is still difficult. In our own species this genus ranges throughout northern Europe, as high as Lapland; a cause for which we shall discover when we trace the geography of the next genus, *Bombus*. One species different from any of ours occurs in the Brazils, and others are found in the Polish Ukraine, and in the United States of North America. The genus appears extremely limited in numbers, for although nearly a hundred of the genus *Bombus* are known, *Apathus*, in collections, seems limited to ten. This may perhaps arise from want of due observation or from the neglect of their careful separation from that genus, but our own species are far from co-extensive with our native species of *Bombus*.

The genus BOMBUS, although with some southern irrepressible propensities, it being found within the tropics in a few instances, is essentially a northern form, which is strongly indicated in its downy habiliments, for it is clothed in fur like the Czar in his costly blue-fox mantle. In the Old World its range extends to Lapland, whither it is followed, as previously noticed, by its parasite *Apathus*, and in the New World to Greenland, where one species seems an autochthon, perhaps originating there when the land was still verdant, and grew grapes, long before the age of Madoc. Other species occur far away to the north of east, booming through the desolate wilds of Kamtchatka, having been found at Sitka; and their cheerful hum is heard within the Arctic circle, as high as Boothia Felix, thus more northerly than the seventieth parallel. They may, perhaps, with their music often convey to the broken-hearted and lonely exile in Siberia, the momentarily cheering reminiscence of joyful youth, and by this bright and brief interruption break the monotonous and painful dullness of his existence, recalling the happier days of yore: but the flowers of humanity, here typified by the natural flowers which attract these stray comforters, will one day spring where the salt of tears now desolates, and thus the merry bees have sweetness for even these poor outcasts, and froth their bitter cup with bubbling hope.

In the south of Europe the genus occurs in Austria, the island of Zante, and the Pyrenees. It is found in Syria, the island of Java, in China at Chusan and Silhet, and also in northern India; and, although crossing the tropics to fix itself at Monte Video, at the mouth of Rio de la Plata, in Africa it appears to be found at Oran only; nor does it occur in Australasia. In South America it is also found at Pará and Cayenne, and on the opposite side at Columbia, Quito, and Chili, and passes up the isthmus to California, and thence to Mexico, whence it extends to the island of Antigua.

The genus APIS, or the HIVE BEE,—which perhaps in its past and present utility to man, may successfully compete in the aggregate with the silkworm,—with true regal dignity comes the last of the series of genera. The whole array of her precursors, who marshal her way, and derive their significance and importance from the more or less direct resemblance in structure and function to her, deduce their common name of "Bees" from this relationship, and consequently from her. Long before their existence had been traced by the observer of nature or by the naturalist, the comb of the BEE had dropped in exuberant luxuriance its golden stores for the gratification of mankind. This little creature had garnered, from sources inaccessible to man, the luscious nectar concealed within the bosom of the flower, whose exquisitely beautiful varieties, in form, colour, and fragrance, had delighted his sight and his smell long before he had been led by accident to discover that these industrious little workers collected into their treasury, from those same flowers, as exquisite a luxury for his taste, as they themselves had yielded to his other senses. Thus the earliest records speak of honey, and of bees, and of wax; and the land of promise to the restored Israelites, was to be a land flowing with milk and honey.

Réaumur, whose observations upon bees had been pursued with such patient and indefatigable perseverance, combined with such minute accuracy, and then recorded so agreeably, and who conceived the possibility of establishing a standard of length, for the common use of all nations, to be derived from the length of a certain number of the honey-cells of the comb, to which notion he was doubtless led by their mathematical precision and uniform exactitude, appears to have been unaware of the existence of other species of the genus, and hence he assumed, in his ignorance of this fact, that in all countries they were alike.

Travellers had, even for more than a century before, mentioned different kinds of honey, derived from different kinds of bees, which, however, Réaumur does not, from this circumstance, seem to have known. Had he been acquainted with it, his philosophical accuracy of observation and habit of reflection would certainly have assumed the possibility of differences of size in the cells of the different bees, and he would have waited until opportunity had given him the power of determining whether this mode of admeasurement could be safely adopted as certainly being of universal prevalence. It is to be wondered at also, that he did not weigh the possibility that climatic differences in the distribution of even the *Apis mellifica* might have involved discrepancies, by the effects constantly seen to be produced by climate, and which would have shown that the standard which he sought to establish could not be relied on.

Collections exhibit about sixteen species of the genus *Apis*, whose natural occurrence is restricted to the Old World, for although the genus, especially in the species *A. mellifica*, has been naturalized in America, and also in Australasia, and in some of the Islands of the Pacific, these were originally conveyed thither by Europeans. Those countries possess representatives of the genus with analogous attributes and functions, in two other genera, which fulfil the same uses. It is remarkable that the Red Indians used to note the gradual absorption of their territory by the White Man, through the forward advance of his herald *Apis mellifica*. This species has also been carried to India, to the Isle of Timor, and to northern, western, and southern Africa, in all which countries it is thoroughly naturalized, although they all possess indigenous species, which are quite as, or perhaps more largely, tributary to their inhabitants. Observation has not hitherto confirmed the identity of the manners of these exotic species with our own, owing to the deficiency of observers with the enthusiasm requisite to follow their peculiarities with the patience of a Réaumur, a Bonnet, or a Huber. That they are quite or all but similar, exclusively of differences of size, both in their habits and their nests, may be inferred from their identity of structure. We know that they consist of three kinds of individuals—neuters, females, and males,—and that their combs are made in cakes built vertically, formed of hexagonal contiguous cells, which are placed bottom to bottom, and overlap each other in the same strengthening position as do ours; and also that the cells wherein the males are developed are oval, larger than the honey-cells, and less uniform. With all these similitudes it is fair to suppose that their economy may be the same; but their honey-cells, from their smaller size, (the bee which produces them being smaller,) have a more elegant appearance; and it is concluded from the largeness of the nest, taken conjunctively with the smallness of the cells, and of the bees constructing it, that the communities thus associated must in their collective number be considerably larger than those of our hives.

Instinct, as expressed in the habits, is as sure a line of separation, or means of combination, as structure, and is corroborative in tending to preserve generic conjunction in its inviolability. And, conversely, with certainty, is indicated that such-and-such a form, in the broad and most distinguishing features of its economy, is essentially the same in every climate. The habits, therefore, in whatever country the genus may occur, may be as surely affirmed of the species, from the knowledge we have of those at home, as if observation had industriously tracked them. This is especially the case in a genus, the species of which present such a peculiar identity of *structure* as does *Apis*, whose specific differences are derived only from colour and size, and this identity is a peculiarity, so far as I have observed, rarely found in other genera, numbering even no more species, but wherein slight differences of structure often yield a subsidiary specific character, complete structural identity being almost solely incidental to the genus *Apis*.

The importance of honey and wax throughout the world, as well for the ceremonies of religion, as for the service of the arts, and for medical or domestic purposes, is attested by the vigilance, care, and assiduity with which bees are tended in every country. Although sugar, since its introduction to those northern countries which have not been favoured by nature with the cane that yields it, has superseded for ordinary uses the produce of the hive, this still continues serviceable for many purposes to which sugar cannot be applied. It is used in many ways in pharmacy, and still retains in the interior of some continents, owing to the deficiency of sugar, arising from the difficulties and expenses of transit, all its primitive uses. In the East, even in countries producing sugar in abundance, honey is extensively employed for the preservation of fruits, which in their ripe state in those hot climates would rapidly lose their fulness of flavour were they not thus protected, — honey here being esteemed superior to sugar in the circumstance of its not crystallizing by reason of the heat, and also from its applicability to this use in its natural state.

This is especially the case in China, where a conserve of green ginger, and of a fragrant orange (the *Cum Quat*), are in high repute, and which are peculiarly grateful to Europeans on the spot. These, however, are so delicately susceptible of change of climate, that they lose some of the aroma that constitutes much of their attraction, upon transportation, and, indeed, like many kinds of Southern wines, can be appreciated only within their own country, from their extreme delicacy and tendency to spoil.

Honey is a very favourite food and medicine with the Bedouins in Northern Arabia. Bees make their hives in all the crevices of rocks in Hedscha, finding everywhere aromatic plants and flowers. At Taif, bees yield most excellent honey, and the honey at Mecca is exquisite. At Veit-el-Fakeh, wax from the mountainous country of Yemen is exchanged for European goods and for spices from the further Indies. In Syria and Palestine we find bees abound. At Ladakiah there are large exports both of honey and wax; and the honey of Ainnete, on the declivities of the Lebanon, is considered the finest of the whole of that mountain-range. Antonine the Martyr, in the seventh century, speaks of the honey of Nazareth being most excellent, and in the present day bees are extensively cultivated at Bethlehem, for the sake of the profit derived from the wax tapers supplied to the pilgrims. Some of the members of the German colony at Wadi Urtas speak of the purchase of eleven beehives at this place, and express themselves as very sanguine of an abundant harvest from the luxuriance and profusion of flowers, although they say the bees are smaller than those of Westphalia, and are of a yellowish-brown colour. The eastern side of this peninsula, especially the district of Oman, is wholly destitute of bees, contrasting thus unfavourably with its western fertility.

The enormous quantities of honey produced may be comparatively estimated by the collateral production of beeswax, which it exceeds by at least ten to one. When we reflect upon what masses of the latter are consumed in the rites of the Roman Catholic and Greek churches throughout the many and large countries where those religions prevail, we shall be able to form a general estimate of the extensiveness and universality of the cultivation of bees. Nor are those the only uses to which wax is applied, and the collective computation of its consumption will show that bees abound in numbers almost transcending belief.

The name of *bougie* for wax-candle or taper, is used

by all the languages of the south of Europe, and is derived from the name of Bugia, a town of Northern Africa, whence, even as long back as the time of the Roman Empire, wax was obtained to make candles for lighting. The inhabitants of Trebizonde paid their tribute to the Roman Empire in wax. Both honey and wax are largely employed in pharmacy, and were also, in ancient times, both extensively used in embalming. The honey of Mount Hymetta in Attica, and of Hybla in Sicily, were each in as high repute in classical countries as is that of Narbonne in Languedoc, by reason of its choice delicacy, with us, and throughout France. Distributed over the wide pastures of the Ukraine, every peasant has his store of hives, which frequently, in their harvests, realize more largely than their crops of grain, —multitudes of that peasantry computing as important items in the estimate of their wealth the number of their beehives, which often exceed five hundred to the individual possessor. In Spain and Italy bees are largely cultivated; and in the former country many a poor parish priest, the religious monitor of an obscure hamlet, can count his five thousand.

In countries so rich in the productions of Flora, whose seasons there are perennial, and which fluctuate only in special locality, bees are removed to and fro to meet these peculiarities. Thus in the south of France, where large tracts are cultivated with aromatic shrubs and flowers, for the distillation of essential oils and fragrant waters, the hives of bees are moved up and down the adjacent rivers upon rafts, as the flowering of the crops succeed each other. In Italy, Spain, and Southern Russia, the same practices are pursued, although we have no detailed accounts of the precise spots; but we know from Niebuhr, Savigny, and Sir Gardiner Wilkinson, that upon the Nile it is customary thus to transport the bees from flower-region to flower-region upon rafts containing about four thousand hives, each numbered by the proprietors of the hives for identification, who thus double the seasons by continually shifting their bees from Lower Egypt to the Upper Nile and back again.

In ancient Greece also, they were conveyed for this purpose from Achaia to Attica; in the former of these provinces, owing to its higher temperature, flowers had passed their bloom before spring had opened in the latter. All these circumstances tend to show that the experience of bee-masters, both ancient and modern, has ascertained that their insects have not a very extensive range of flight.

Of the fact that the honey of bees is not always salutary to man, there is a remarkable instance recorded in Xenophon, in his narrative of the retreat of "The Ten Thousand," who reports that upon falling in with quantities of it, in Asia Minor, those who indulged in its enjoyment were seized with vertigo, or headache, and violent diarrhœa, attended with sickness, but which had no fatal consequences, although they did not recover from its injurious effects for a couple of days, and were left then in a very prostrated condition. The celebrated physician and botanist Tournefort, when travelling in the East, towards the end of the seventeenth century, found, in the neighbourhood of Trebizonde, an excessive luxuriance of the flowers of the *Rhododendron ponticum* and of the *Azalea pontica*, which, although sumptuous in their blossoms, were held in bad repute by the inhabitants, who ascribed to their odour the deleterious effect of causing headache and vertigo. He was thence induced to surmise that these had possibly been the flowers the bees had extracted the honey from which had been so baneful to the troops of Xenophon.

But it seems that bees themselves cannot collect with impunity the honey of noxious flowers, for they are occasionally subject to a disease resembling vertigo, from which they do not recover, and which is attributed to the poisonous nature of the flowers they have been recently visiting.

Several different kinds of honey and wax have been described, but some degree of uncertainty exists as to whether they are all the produce of genuine species of the genus *Apis;* for it will be found, in a rapid notice I purpose giving of the more conspicuous genera of foreign bees, that there are two exotic genera of this section of the family, both social in their habits, and which both produce the same materials; there is a wasp also that makes honey. But of all the many kinds of honey noticed, the green kind furnished to Western India by the island of Réunion, the produce of an *Apis* indigenous to Madagascar, but which has been naturalized in the French island, and also in the Mauritius, is perhaps the most remarkable. It is of a thick syrupy consistency, and has a peculiar aroma. It is much esteemed upon the most proximate coasts of the peninsula of India, where it bears a high price. Whether its greenness of colour is derived from the flowers which this species frequents, or whether it be incidental to the nature of the bee, has not been ascertained, but the honey of the South American wasp, the sole species producing the material, has also a green tinge.

Nature has assigned the task of thus catering for man, by collecting and garnering from the recondite crypts within the blossoms of flowers, to about sixteen species congenerical with our honey-bee, but sufficiently differing. As I have before noticed, the species of this genus greatly more resemble each other in structure than perhaps do the species collocated within any other genus of insects, and whence may be inferred an exact similitude of habits, although as yet unconfirmed by direct observation.

The second European species, the *Apis Ligustica*, or Ligurian bee, is rather larger, but very like ours, and inhabits the whole of the north of Italy, its occupation of that country extending from Genoa to the vicinity of Trieste; its progress further north being impeded by the Alps of Switzerland and the Tyrol. It is also found in Naples, and may likewise spread to the Morea, Turkey, and the Archipelago of Greece, and is perhaps the bee noticed by Virgil. Either this species, or possibly one distinct from ours, is that which is so extensively cultivated in Spain, although ours is found in Barbary.

Another smaller kind, the *Apis fasciata*, has been cultivated in Egypt from time immemorial, and which yielded its abundant harvests for the gratification of the ancient Romans. Only five other distinct species, so far as is yet known to us, appear to occupy the vast continent of Africa,—two on its western coast at Senegal and Congo, the *A. Adansonii* and the *A. Nigritarium*; two in Caffraria, the *A. scutellata* and the *Apis Caffra*. That at Madagascar, and doubtless on the adjacent mainland, which has also been naturalized in the Mauritius and at Réunion, is the *Apis unicolor*, which produces the green honey mentioned above.

India, however, at present appears to be the true metropolis of the genus. Further discoveries in Africa may hereafter give that vastly larger continent the predominancy;

but there is no doubt that, so far as present information
extends, India has the superiority. Thus *Apis dorsata*,
Apis nigripennis, and *Apis socialis*, are cultivated in
Bengal, the latter being also found along the Malabar
coast and at Java. It is singular that the only instance
of the occurrence of the very distinct genera of *Apis*
and *Mellipona*, both honey-storing genera, yet known
to exist indigenously in the same locality, is found in
this island. At Pondicherry and its vicinity are found
Apis Delessertii and *Apis Indica*. This latter bee is
extensively cultivated, and its hives are perhaps the
most largely inhabited of any of the species; the numbers occupying a single nest being estimated at above
eighty thousand.

From India also, but to which no special locality is
assigned, come *Apis Perrottetii*, *Apis lobata*, as likewise
Apis Peronii, which is equally native to the Isle of
Timor. The honey produced by this last bee is yellow,
more liquid than ours, and of a very agreeable flavour.

Thus science dissipates the popular supposition, that
a multiplicity of the individuals of one species of this
insect produces the tons of wax and the myriads of
gallons of honey that are annually consumed.

Which of these bees first benefited the human race,
in its primitive seat, and before the multiplication of
mankind forced them to take divergent courses from the
cradle of their birthrace, "to people the whole earth," it
is impossible to say. And it is equally impossible to conjecture whether, like man, they by this course of migration have assumed the features they now exhibit of distinctly different species; yet they do not vary so considerably among themselves as do many other creatures that
have come under the direct influence of man,—the chief

differences consisting in the comparatively slight distinctions of colour and of size, but which are sufficiently marked to constitute them good species.

The earliest manuscript extant, which is the Medical papyrus, now in the Royal Collection at Berlin, and of which Brugsch * has given a facsimile and a translation, dates from the nineteenth or twentieth Egyptian dynasty, accordingly from the reign of Ramses II., and thus goes back to the fourteenth century before our era. But a portion of this papyrus indicates a much higher antiquity, extending as far back as the period of the sovereigns who built the Pyramids, consequently to the very earliest period of the history of the world.

It was one of the medical treatises contained within the Temple of Ptah, at Memphis, and which the Egyptian physicians were required to use in the practice of their profession, and if they neglected such use, they became responsible for the death of such patients who succumbed under their treatment, it being attributed to their contravening the sacred prescriptions. This pharmacopœia enumerates amongst its many ingredients, honey, wine, and milk; we have thus extremely early positive evidence of the cultivation of bees. That they had been domesticated for use in those remote times, is further shown by the fact mentioned by Sir Gardiner Wilkinson of a hive being represented upon an ancient tomb at Thebes.

It may have been in consequence of some traditional knowledge of the ancient medical practice of the Egyptians, that Mahomet, in his Koran, prescribes honey as a medicine. One of the Suras, or chapters, of that

* 'Recueil de Monuments Égyptiens dessinés sur les lieux.' In Three Parts. 4to. Leipzig, 1862.

work, is entitled 'The Bee,' and in which Mahomet says:—" The Lord spake by inspiration unto the Bee, saying, 'Provide thee houses in the mountains and in the trees [clearly signifying the cavities in rocks and hollows of trees, wherein the bees construct their combs], and of those materials wherewith men build hives for thee; then eat of every kind of fruit, and walk in the beaten paths of thy Lord.' There proceedeth from their bellies a liquor of various colours, wherein is a medicine for men. Verily herein is a sign unto people who consider."

It is remarkable that the bee is the only creature that Mahomet assumes the Almighty to have directly addressed. Al-Beidawi, the Arabic commentator upon the Koran, whose authority ranks very high, in notes upon passages of the preceding extract, says, "The houses alluded to are the combs, whose beautiful workmanship and admirable contrivance no geometrician can excel." The "beaten paths of thy Lord," he says, "are the ways through which, by God's power, the bitter flowers, passing the bee's stomach, become honey; or, the methods of making honey he has taught her by instinct; or else the ready way home from the distant places to which that insect flies." The liquor proceeding from their bellies, Al-Beidawi says, "is the honey, the colour of which is very different, occasioned by the different plants on which the bees feed; some being white, some yellow, some red, and some black." He appends a note to where Mahomet says, "therein is a medicine for man," which contains a curious anecdote. The note says, "The same being not only good food, but a useful remedy in several distempers. There is a story that a man once came to Mahomet, and told him his brother was afflicted with a violent pain in his belly; upon which the Prophet bade him give him some honey. The fellow took his advice; but soon after, coming again, told him that the medicine had done his brother no manner of service. Mahomet answered: '*Go and give him more honey,* for God speaks truth, and thy brother's belly lies.' And the dose being repeated, the man, by God's mercy, was immediately cured."

That the primitive Egyptians were familiar with the peculiar economy of the bee in its monarchical institution is proved by the figure of the bee being adopted as the symbolical character expressive of the idea of a people governed by a sovereign This figure is frequently met with upon Egyptian sculptures and tablets, dating as far back as the twelfth dynasty; but upon these the bee is very rudely represented, being figured with only four legs and two wings; but upon a tablet of the twentieth dynasty the bee is correctly represented with four wings and six legs.

All these facts take us far back in the history of the bee. But the indication of a higher antiquity of its domestication may be traced in the Sanskrit, wherein *ma* signifies honey, *madhupa,* honey-drinker, and *madhukara,* honey-maker, the root of the latter signifying "to build." *Madhu* has clearly the signification of our *mead,* thence we may thus trace an affinity, pointing to those early times, for the origin of a drink still in use amongst us. In Chinese *mih,* or *mat* (in different dialects) signifies honey, thus clearly showing a second derivation, in this Turonian term, from a more primitive language whence both flowed. In the Shemitic branch nothing analogous is to be traced. But this double convergence to a more distant point veiled in the obscurity of time, necessarily takes the domestication of the bee back also to that anterior period now only dimly traceable.

There can be but little doubt that the majority of the creatures now domesticated by man were in those ancient days subjected to his sway, and to which later times have not added any, or but few fresh ones. A natural instinct possibly prompted him originally in the selection; and if the reindeer of the Laplander seem an aberration, this has happened through the contingency of climate, for in the high latitudes it inhabits, it, in its uses to man, supplies the double function performed in more southern regions by the equine and bovine tribes.

In the Greek and in the Teutonic languages, two branches of the Aryan stem, the names of the bee, *melissa* and *biene*, are clearly derived from the constructive faculty of the insect, and to which the root of the Sanskrit word *madhukara*, above noticed, also points. It would seem, therefore, that an earlier notice of its skill than of its honey, had suggested its name. Thus everything points to a very early acquaintance with the bee, its economy, and its properties, and this familiarity might be easily traced down in regular succession to the present times, were it desirable to recapitulate what has been so often repeated in the history of the " Honey-bee." The facts I have gathered together above, do not seem to have been hitherto strung together, and may be suggestive of reflection, as well as affording some amusement.

The study of the geographical distribution of natural objects has a more universal bearing, and yields collectively more definite instruction and information than its partial treatment, when restricted to small groups, may at first seem to promise. This, however, is very useful, for it is but by the combination of such special details that the enlarged views are to be obtained, from which theories of the general laws of distribution can be deduced. Of course, small creatures with locomotive capacities will not supply the positive conclusions that may be framed from such objects as are fixed to their abode, and have not the same power of diffusion, although they certainly appear to be generally restrained within particular limits by physical conditions of the earth's surface subservient to the maintenance of special forms of organic life; and these, once determined, would yield and derive reciprocal illustration. They may be merely climatic, but climate thus indicated cannot be estimated by zones, or belts, or regions; for they seem to traverse all these, and follow undulations not specially appreciable except in the results they exhibit.

Unfortunately the bees have been too imperfectly collected, and too irregularly registered, to admit of arriving at any precise conclusions with respect to them. All that can as yet be done will be to combine the scanty notices afforded by the contents of our collections, in the hope that their promulgation may induce collectors, who happen to have the often extremely rare opportunity of examining distant countries, to avail themselves of the happy chance, which may never recur, or only at long intervals.

Nor can I too impressively reiterate the importance of noting both special localities, altitude, temperature, season, flora, etc., as being all conducive to the widest instruction upon the subject. Indulging in the hope that travellers will act upon these suggestions, and thus considerably add to the value of what they may industriously collect, we must patiently await until time brings it about.

Encouraging this expectation, I have summarily collected, under their topical arrangement, the notices which precede, but which are there arranged in the generic order of the bees.

From the information we thus possess, we learn that some of our genera have an extremely wide diffusion, and occur in countries where we might have expected that other forms would have superseded them in the offices they are ordained to fulfil. None of the schemes for the geographical distribution of insects yet propounded, seem to curb the eccentricities of their range. The regions proposed by Fabricius in his 'Philosophia Entomologica,' they break through as readily as through the concentric circles of the cobweb when this opposes them: and all I can do is to present them as they offer themselves, with the remark that the occurrence of solitary forms in certain localities are almost sure indications that allied genera would be found at hand were they heedfully sought. It will also be observed, that in some places a parasitical genus, and its known sitos, only, have been captured there.

The following list will strongly show how totally our genera of bees are unaffected by isothermal, isotheral, or isochcimal lines drawn over the earth's surface. Nor do botanical conditions seem to influence them beyond the probability of their dissemination being restricted to the special diffusion of the families of such plants whose genera and species they frequent with us.

Thus, inhabiting Northern Europe we find in—

Lapland. Cilissa; Anthophora; Epeolus; Megachile; Chelostoma; Heriades; Osmia; Apathus; Bombus; Apis.

Finland. Colletes; Prosopis; Cilissa; Anthophora; Nomada; Epeolus; Stelis; Cœlioxys; Megachile; Anthidium; Chelostoma; Heriades; Osmia; Apathus; Bombus; Apis.

Sweden. All our genera except Sphecodes; Halictus; Macropis; Anthocopa.

Denmark. All our genera except Macropis and Anthocopa.

Russia. All our genera except Macropis and Anthocopa.

The other Northern European Countries. All our genera, with the same exceptions.

Western, Southern, and Eastern Europe present us with, in—

France. All our genera.

Portugal. Prosopis; Sphecodes; Andrena; Halictus; Eucera; Nomada; Anthidium; Apathus; Bombus; Apis.

Spain. Prosopis; Sphecodes; Andrena; Halictus; Dasypoda; Eucera; Anthophora; Nomada; Megachile; Anthidium; Apathus; Bombus; Apis.

Italy. Andrena; Halictus; Panurgus; Eucera; Anthophora; Nomada; Melecta; Epeolus; Cœlioxys; Megachile; Anthidium; Osmia; Apathus; Bombus; Apis.

Sicily. Prosopis; Sphecodes; Eucera; Anthophora; Melecta; Epeolus; Megachile; Anthidium; Osmia; Apathus; Bombus; Apis.
Malta. Halictus; Apis.
Isles of Greece. Dasypoda; Apis.
The Morea. Prosopis; Sphecodes; Halictus; Dasypoda; Eucera; Anthophora; Ceratina; Nomada; Melecta; Anthidium; Chelostoma; Osmia; Bombus; Apis.
Albania. Prosopis; Sphecodes; Dasypoda; Eucera; Ceratina; Nomada; Melecta; Megachile; Anthidium; Osmia; Bombus; Apis.
Dalmatia. Halictus; Eucera; Anthophora; Megachile; Anthidium; Apis.

Asia exhibits to us, in—
Siberia. Andrena; Nomada; Epeolus; Bombus; Apis.
Kamchatka. Bombus.
China. Halictus; Nomada; Anthophora; Megachile; Bombus; Apis.
Northern India. Prosopis; Sphecodes; Andrena; Halictus; Ceratina; Nomada; Cœlioxys; Megachile; Bombus; Apis.
Bengal. Anthophora; Ceratina; Apis.
Tranquebar. Nomada; Apis.
Ceylon. Anthophora; Ceratina; Apis.
Bombay. Anthophora; Megachile; Apis.
Arabia Felix. Anthophora; Anthidium; Apis.
NOTE.—The genus *Apis* does not occur in *Oman.*
Mesopotamia. Eucera; Nomada; Melecta; Megachile; Anthidium.
Syria. Halictus; Eucera; Anthophora; Cœlioxys; Anthidium; Bombus; Apis.

H

In Africa we find, in—

Egypt. Colletes; Sphecodes; Andrena; Dasypoda; Eucera; Anthophora; Saropoda; Cœlioxys; Anthidium; Osmia; Apis.

Nubia. Anthidium; Anthophora; Apis.

Abyssinia. Megachile; Apis.

Tunis. Dasypoda; Nomada; Apis.

Algeria. Colletes; Prosopis; Sphecodes; Andrena; Panurgus; Eucera; Anthophora; Ceratina; Nomada; Cœlioxys; Megachile; Anthidium; Osmia; Bombus; Apis.

Barbary. Halictus; Nomada; Anthidium; Osmia; Apis.

Madeira. Halictus; Apis.

Canaries. Colletes; Sphecodes; Andrena; Halictus; Anthophora; Melecta; Osmia; Apis.

Senegal. Halictus; Anthophora; Ceratina; Megachile; Apis.

Gambia. Cœlioxys; Megachile; Anthidium; Apis.

Sierra Leone. Halictus; Cœlioxys; Megachile; Anthidium; Apis.

Coast of Guinea. Anthophora; Cœlioxys; Megachile; Anthidium; Apis.

Fernando Po. Megachile.

Western Africa. Halictus; Apis.

Cape of Good Hope. Halictus; Anthophora; Ceratina; Epeolus; Cœlioxys; Megachile; Anthidium; Apis.

South Africa [no distinct locality]. Halictus; Saropoda; Apis.

Natal. Anthophora; Cœlioxys; Megachile; Anthidium; Osmia; Apis.

Madagascar. Apis.

Réunion. Halictus; Apis.

Mauritius. Megachile; Apis.

In America we find, in—
Arctic America and Hudson's Bay. Prosopis; Andrena; Halictus; Megachile; Osmia; Bombus.
Canada and Nova Scotia. Andrena; Halictus; Nomada; Cœlioxys; Megachile; Osmia; Bombus.
United States. Colletes; Sphecodes; Andrena; Cilissa; Halictus; Eucera; Anthophora; Ceratina; Epeolus; Stelis; Cœlioxys; Anthidium; Chelostoma; Heriades; Osmia; Apathus; Bombus.
Mexico. Anthophora; Epeolus; Megachile; Anthidium; Bombus.
California. Bombus.
Columbia. Colletes; Bombus.
Quito. Bombus.
Chili. Sphecodes; Halictus; Anthophora; Melecta; Epeolus; Anthidium; Bombus.
Jamaica. Halictus.
Cuba. Anthophora; Cœlioxys; Megachile.
St. Domingo. Anthophora.
Antigua. Bombus.
Guadeloupe. Anthophora.
St. Thomas's. Cœlioxys; Megachile.
St. Croix. Megachile.
Cayenne. Halictus; Eucera; Ceratina; Cœlioxys; Anthidium; Bombus.
Pará. Anthophora; Cœlioxys; Bombus.
Brazils. Prosopis; Halictus; Ceratina; Epeolus; Stelis; Cœlioxys; Megachile; Anthidium; Apathus; Bombus.
Paraguay. Anthophora.
Monte Video. Bombus.

In Polynesia there occur—
Sandwich Islands. Prosopis.
Philippines. Anthophora; Nomada; Megachile.

In Australia are found—
Swan River. Prosopis; Megachile.
Adelaide. Prosopis; Megachile.
Port Phillip. Prosopis.
Tasmania. Prosopis; Megachile.
Sydney. Sphecodes; Halictus.
New Zealand. Halictus.
Australia [but no distinct locality]. Anthophora; Saropoda.

CHAPTER IV.

NOTICE OF THE MORE CONSPICUOUS FOREIGN GENERA OF BEES.

SEEING thus the wide and almost universal distribution of many of our own genera, we might be induced to ask whether this could not suffice, by the impetus which more genial climates give to the multiplication of individuals, to meet all the exigencies of the most favoured regions of the vegetable kingdom. This is not so. There seems scarcely a limit to the exuberance wherein nature revels in the production of variations of form. The splendour, elegance, and infinite variety which she displays in her floral beauties in the most luxuriant climates, find rivalry as well in the multitude as in the magnificence of the insects which she has allied with them as the indispensable promoters of their perpetuation. How otherwise than through some of the insects we shall mention could tropical *Labiatæ* and the tubulated flowers of the *Rubiaceæ*, etc. be fertilized? The reader will therefore, I trust, welcome an acquaintance with some of the most conspicuous of the group of bees produced by tropical countries, although the main object of this treatise is to exhibit the attractions of "our native bees."

I will but superficially and rapidly glance at the

more distinguished exotic genera and species, as supplementary to the preceding notice of the geographical range of those which are indigenous with us.

How our own species reached us is a subject which has at present eluded all satisfactory determination. For its solution we must await the further discoveries of geology; at present we can only attribute their advent here to the same causes which are common to the production of all our groups of both the animal and the vegetable kingdoms.

Knowing how affluent tropical and subtropical countries are in the variety, size, and number of the forms, as well as in the splendour of their plants and vertebrated animals, we may fairly expect as gorgeous a richness in the insects they produce. Nor shall we be disappointed, for the imperial magnificence of their *Lepidoptera* and *Coleoptera* guarantees an equivalent brilliancy in the other orders of insects, and which is fully confirmed by the harmonious splendour of their bees.

They thus put forward claims to attention and must excite curiosity by their beauty and size, which the comparative smallness of our own, and the usual dulness of their colours do not possess. The latter only repay notice upon close investigation, but they then as amply reward all labour bestowed upon them by the mental recreation they yield, as their more gaudy exotic rivals. The former present themselves obtrusively and exact notice, whereas ours meekly solicit it by their humble but solid allurements. Here, as well as there, we behold the works of a mighty hand and of an immeasurable intelligence.

The bees throughout the world, as known collectively to the richest cabinets, number about two thousand species. This host, in itself numerically so large, solicits attention, for it is opposed to the economy of nature that there should exist any without functions of essential usefulness, making them important elements in her harmonious order and necessary to her due course, irrespective of the instruction to be derived from the study of the manifold varieties of structure, which unquestionably point to distinguishing peculiarities of habits.

In the true bees the division of the *Dasygasters* presents the fewest differing generic forms: the *Nudipedes* and *Scopulipedes* exhibit more numerous varieties, the preponderance being in favour of the pollen-collecting bees (the latter), although the cuckoo bees (the *Nudipedes*) are very abundant, and taken *en masse*, are certainly the handsomest. If it be absolutely the case that there are no parasites amongst the *Andrenidæ*, this subfamily will add very largely to the exotic pollinigerous majority, which thereby becomes extensively subservient to the fruition of the vegetable kingdom.

Those bees which are exclusively inter- or sub-tropical, seem furnished with larger capacities for fulfilling the special mission to which the family is appointed. Their pollinigerous and honey-collecting organs are peculiarly adapted both to the structure and luxuriance of the superb vegetation of those regions, and to which they seem distinctly limited. But that they are not considered equivalent to the entire demand of the profuse bloom everywhere abounding, may be concluded from the tropical range and distribution of many of our northern forms. Thus, whilst the flora of those climates is strictly circumscribed in its diffusion, its fauna, distinctly in the class of insects, and especially in the family of bees, is very considerably less limited in extension.

The exotic genera of bees which are peculiarly noticeable, either from splendour, size, or remarkable eccentricities of structure, are numerous. Tropical and subtropical regions of course abound with them, in individuals, in species, and in genera; and when we reflect upon the riches of the flora of those countries, which is perpetuated mainly by the agency of insects, amongst which, in fulfilling this indispensable demand, bees, as I have reiterated, are pre-eminently conspicuous, we shall not even wonder that their number, although excessive in the extreme, is considerably aided, in many cases, in the performance of this task, by peculiarities of structure. Thus, the splendid Brazilian genus *Euglossa*, although not conspicuous for size, is remarkably so for the enormous development of its posterior tibiæ, which form very large triangles, compared with the size of the insect, deeply hollowed for the conveyance of pollen. Its tongue also, from the length of which the genus derives its name, is, when extended, more than twice the length of the body, and with which it is enabled to reach the nectarium, seated within the depths of the longest tubes of flowers. Other exotic bees, further to aid them in collecting pollen, in addition to the dense brushes with which their posterior legs are variously covered, have each individual hair of these thick brushes considerably thickened by hairs given off laterally, and in some cases these again ramify. Sometimes, in variation, the simple, single hairs have a spiral curve, which almost equally enlarges the activity of their operation. This is also the case with two very hairy-legged genera of our native bees, proximately allied to each other in the methodical arrangement, *Dasypoda* and *Panurgus*, the hair of whose posterior legs have this spiral twist. The most hairy-legged exotic bees are essentially the genera *Centris* and

Xylocopa. Of the habits of the former we know nothing, but those of the latter we are intimately acquainted with, through the elaborate descriptions given by Réaumur and the Rev. L. Guilding, the latter of whom made his observations upon a species found in the island of St. Vincent's, in the West Indies. This last genus exhibits in some of its species the giants among the bees, and one is especially so, a native of India, the *Xylocopa latipes*, which is an inch and a quarter long, and more than three inches in the expansion of its black, acute wings; and it is also noticeable from the anterior tarsus in the male being greatly dilated and white, the bee itself being intensely black, and which in this same sex has enormous eyes united at the vertex, as in the male *Apis*, or drone. In this genus, as in many other genera of bees, there is often a great discrepancy in the appearance of the sexes, they being so totally dissimilar that no scientific skill has hitherto been able to discover a clue for uniting together correctly, by scientific process merely, the sexes of a species; thence the numbers of the species in such genera are unduly augmented beyond their natural limits, from the fact of observation having neglected to associate the legitimate partners.

In some of our native genera this same difficulty existed, which, however, is gradually diminishing as the authentic sexes are slowly discovered.

Exotic bees exhibit also a peculiarity I had occasion to observe before, in reference to our own bees, amounting perhaps to a law, viz the more highly-coloured condition of the parasite, for we find all the parasitical bees of those latitudes, usually gorgeously arrayed in metallic splendour, as instanced in *Aglaë*, *Mesonychia*, *Mesocheira*, etc., and *Melissoda* (my *Ischnocera*, in Lardner), is remarkably conspicuous for its long and delicately slender antennæ in the male, each joint of which is nodose at its extremity.

The widely-distributed *Nomia* seems to abound chiefly in India. It, although neither gay nor large, has, in its males, a distinguishing form of the posterior tibiæ, which is greatly incrassated or thickened; a peculiarity of structure found also in some other genera of *Hymenoptera*, and in several genera of the *Diptera*, giving the insects which have it a remarkable gait.

The singularly anomalous distortion of these posterior legs is conspicuous also in the genus *Ancylosceles*, which is named in allusion to it.

Another remarkable peculiarity is to be observed in the above genus, *Mesocheira*, as likewise in the superb *Acanthopus*, both of which genera have the spur of the intermediate leg palmated at the extremity, and the latter genus is further distinguished by its large size and splendid development, and by having the fifth joint of the tarsus of the posterior legs longer than the three preceding united, and covered with a pollinigerous brush as dense as that of the elongate first joint of the same limb.

But the foreign genera which will be most interesting to the reader will, I expect, be those of *Trigona* and *Mellipona*, which, in many peculiarities, seem abortive *Apes*. They seem nature's first endeavour to construct *Apis*, for they have an apparently imperfect neuration of the wing, in which the external submarginal cell is unfinished. Their only separating distinction from each other is the difference in their mandibles, which in *Mellipona* are broad and edentate, whereas in *Trigona* they are also broad but denticulated. In *Apis* these organs are merely irregularly enlarged at the extremity, and hollowed within, rather like a spoon, which structure would of course imply a difference of economy.

A further characteristic of these genera, and in which they participate with *Apis*, is the deficiency of spurs to the posterior tibiæ, which separates them from all other genera of bees, as also from *Bombus*, which has two, yet with which, in point of their economy, they more closely assimilate than with *Apis*. They are the South American and Australian indigenous representatives of the genus *Apis*, and are found likewise in Java and Sumatra, and in some of the larger and extreme islands of the Indian Archipelago, thus also similarly in countries where marsupial animals occur. Like *Apis*, they are social in their habits; but their neuters only are as yet known, neither males nor females having been described. They are reputed to be stingless, and to make honey and wax in enormous quantities. The combs in *Mellipona* are attached either to the branches of trees or are suspended from them, but how they are enveloped for security is not reported, but sometimes, like *Apis*, they construct them within hollow trees and in the cavities of rocks, as in *Trigona*, in like manner as *Apis* does in its natural state. Their communities are not so large as those of the hive bee, and the cells of their combs are less perfectly hexagonal, the wax being expended upon them in denser quantities, whereas the hive bee is exceedingly parsimonious in the use of this material, a circumstance arising possibly from the different and more difficult mode the latter have of obtaining it. In the latter it is a secretion; but these exotic genera possibly collect their wax ready-made by the exudation of plants, and, thus, having more readily obtained it, they are more lavish in its use.

Early travellers and historians describe many kinds of honey made by these bees, native to the South American continent, but they report nothing of the peculiarities of the social economy of these insects, nor whether they are as closely allied in this respect to *Apis*, as they are in the collection of honey and wax.

To enter into further detail relative to them would be beyond the province of this work, and I have only given this extremely superficial and brief notice of foreign genera, to show what multitudes of others of this interesting family await admiration and study, when some proficiency has been acquired in the knowledge of our own.

CHAPTER V.

PARASITES OF BEES AND THEIR ENEMIES.

Nature seems to have imposed a restraint upon the undue increase of all its creatures, by creating, to check it, others that prey upon them. It thus enlarges the sphere of its activity by making life accessory to life, and promoting thereby a more extended enjoyment of all its pleasures. Other forms are brought into existence, and other terms given to duration than those which the laws of life attach to specific organization. No abatement is thereby made upon the quantity of contemporaneous vitality, for what subsides in one rises in another, and the undulation of the waves is perpetual.

Does the quantity of life, extant upon the earth, vary? Perhaps mortality ever comes in some shape to prevent it, when excess threatens to render its energy effete. Yet under every circumstance the wise arrangements of Providence suffice, for everything has its enemies or its parasites, which are also enemies, but frequently in disguise. For defence there is an implanted instinctive fear, or abhorrence; and the creature is then left to its skill, prudence, or strength, either to evade or to mitigate, to the extent of its capability, the danger of the attack.

We find the bees are not at all exempted from this prevailing condition. They have many enemies and parasites of remarkably differing organization. They are attacked by many kinds of birds, among which the *Merops Apiaster* (or bee-eater) is conspicuous. All the swallow tribe prey upon them, as do the shrikes and some of the soft-billed small birds, and also many small quadrupeds when they can find the opportunity. Wasps also attack them, but they do not often get entangled in spiders' nets, being generally too strong for the retention of its meshes, but I have seen a *Bombus* enveloped in a tangle of its wonderful filament.

The wild bees' parasites are of two kinds, personal, and such which, like the young of cuckoos, live at the expense of the offspring. The personal parasites are again of two kinds, for bees are infested with several kinds of *Acari*, and once I found a *Bombus* upon the ground in Coombe Wood so swarming with the *Acarus* that it lay hopelessly helpless until I threw it into a pool of water, when its *attachés* were washed away. But the poor bee seemed so prostrated by their attack, that even when freed from them it had not energy to fly, and having landed it I left it to the kindly nursing of nature.

A little yellow hexapod larva sometimes also infests the wild bees in great numbers, running over and about them with great activity. I have never followed these to their development, but they are said to be the larvæ of *Meloe proscarabæus*, a conspicuously large coleopterous insect. The assertion has produced much discussion; and I believe the larva has been bred to the imago, and consequently it has been proved that it is the larva of that insect. But that it should be parasitical upon so small a creature, and that numbers should infest it for their nutriment, is extremely improbable. It is far more likely that instinct has taught them to be conveyed elsewhere through the medium of the bee, as they might also be by attaching themselves to any other volatile insect, and that upon arriving at a suitable locality they descend from their temporary hippogriff. We see seeds thus conveyed by the agency of animals and birds to suitable places, where they fall and germinate.

Another little hexapod is occasionally found upon them: this is intensely black, and like the former, very active: these I never could rear, nor did they ever seem to enlarge, and they speedily died. I have found them in profusion also within the flowers of *syngenesious* or composite plants, especially of the dandelion in the spring.

But their most remarkable personal parasites consist of some very extraordinary insects, so anomalous in their structure as to have required the construction of an order for their reception,—the Order *Strepsiptera*, or "twisted-winged," thus named from the twist taken by their anterior wings or wing-cases. Their natural history is but imperfectly known, and I believe the males have not yet been discovered. Their larva lives within the bee, and feeds on its viscera by absorption, being attached within by a sort of umbilical cord. It presently consumes the viscera, and renders the bee abortive, by destroying its ovaries, for it is usually upon female bees that it is found. When full fed it forms a case within which it changes into the pupa and imago, the head of which case protrudes between the scales of one of the dorsal segments of the abdomen. How it becomes deposited within the bee or the bee's larva remains a mystery, although many hypotheses have been hazarded to account for it, but all are unsatisfactory. The Order consists of three genera (*Stylops, Elenchus,* and *Halictophagus*) found in England, and other parts of Europe; indeed, the genus *Elenchus* has been also discovered in the Mauritius. The Continent possesses the genus *Xenos*, of the same order, and parasitical upon a wasp, neither of which occur with us.

Mr. Kirby, in studying the bees for his invaluable 'Monographia Apum Angliæ,' first came across this extraordinary creature. His description of his discovery is highly interesting. He says, at page 111 of volume ii. of the above work, that having observed a protuberance upon the body of the bee, he was anxious to ascertain whether it might be an *Acarus*, and goes on : " What was my astonishment when, upon attempting to disengage it with a pin, I drew forth from the body of the bee, a white fleshy larva, a quarter of an inch long, the head of which I had mistaken for an *Acarus*. How this animal receives its nutriment seems a mystery. Upon examining the head under a strong magnifier, I could not discover any mouth or proboscis with which it might perforate the corneous covering of the abdomen, and so support itself by suction; on the under side of the head, at its junction with the body there was a concavity, but I could observe nothing in this but a uniform unbroken surface. As the body of the animal is inserted in the body of the bee, does that part receive its nutriment from it by absorption? After I had examined one specimen, I attempted to extract a second, and the reader may imagine how greatly my astonishment was increased, when, after I had drawn it out but a little way, I saw its skin burst, and a head as black as ink, with large staring eyes, and antennæ consisting of two branches, break forth, and move itself briskly from side to side. It looked like a little imp of darkness just emerging from the infernal regions. I was impatient to become better acquainted with so singular a creature. When it was completely disengaged, and I had secured it from making its escape, I set myself to examine it as carefully as possible; and I found, after a careful inquiry, that I had not only got a nondescript, but also an insect of a new genus whose very class seemed dubious."

As everything connected with so strange a creature is very attractive, I will cite what other observers also have seen. Mr. Dale, from whom Curtis received *Elenchus* to figure in his 'British Entomology,' vol. v. pl. 226, says: "These parasites look milk-white on the wing, with a jet-black body, and are totally unlike anything else. It flew with an undulating or vacillating motion amongst the young shoots of a quickset hedge, and I could not catch it until it settled upon one, when it ran up and down, its wings in motion, and making a considerable buzz or hum, as loud as a *Sesia*; it twisted about its rather long tail, and turned it up like a *Staphylinus*. I put it under a glass and placed it in the sun; it became quite furious in its confinement, and never ceased running about for two hours. The clytra or processes were kept in quick vibration, as well as the wings; it buzzed against the sides of the glass with its head touching it, and tumbling about on its back. By putting two bees (*Andrena labialis*) under a glass in the sun, two *Stylops* were produced: the bees seemed uneasy, and went up towards them, but evidently with caution, as if to fight; and moving their antennæ towards them, retreated. I once thought the bee attempted to seize it; but the oddest thing was to see the *Stylops* get on the body of the bee and ride about, the latter using every effort to throw his rider.

"As the *Stylops* emerges from the body of the bee, the latter seems to suffer from much irritating excitement."

Mr. Thwaites writes to me, on the 12th May, thus: "I had the good fortune to capture a *Stylops* flying, and on the Tuesday following saw at least twenty flying about in the garden, but so high from the ground that I could capture only about half-a-dozen; since that time they have become gradually more scarce.

"The little animals are exceedingly graceful in their flight, taking long sweeps as if carried along by a gentle breeze, and occasionally hovering at a few inches distance from the ground. Their expanse of wing and mode of flight give them a very different appearance to any other insect on the wing. When captured they are exceedingly active, running up and down the sides of the bottle in which they are confined, moving their wings and antennæ very rapidly. Their term of life seems to be very short, none of those I have captured living beyond five hours, and one I extracted from a bee in the afternoon was dead the next morning.

"All the bees stylopized, both male and female, I have taken, have manifested it by having underneath the fourth (invariably) upper segment of the abdomen a protuberance which is scale-like when the *Stylops* is in the larva state; but which is much larger and more rounded when the *Stylops* is ready to emerge. A bee gives nourishment generally to but one *Stylops*; but I have occasionally found two, and once three larvæ in one bee."

The structure of these insects is very remarkable: the typical genus *Stylops* is named from its compound eyes, which consist of a very few (about fifteen) hexagonal facets, seated upon a sort of footstalk. The mandibles are lancet-shaped and very acute, and the head, by reason of the protuberant eyes, has very much the shape of a dumb-bell. The antennæ are branched, but in *Halictophagus*, they are flabellate. The thorax is greatly developed; the superior wing is like a rudimentary wing-case, and is twisted, the inferior wings are very large, and fold along the abdomen in repose like a fan; the legs are slender, and the tarsi with four joints in *Stylops*, with three in *Halictophagus*, and with two in *Elenchus;* the abdomen is long, very flexible, and consists of eight segments. The insects themselves do not exceed a quarter of an inch in length in the largest, but they are generally very much smaller. The perfect insect is very short-lived, not surviving many hours, as just stated. They are usually found in the months of May and June, and they have been discovered to infest several species of *Andrena* and *Halictus*, for instance the *A. nigro-ænea*, upon which Mr. Kirby first found it; *A. labialis*, which I have frequently caught stylopized; *A. rufitarsis, fulvicrus, Mouffetella, tibialis, Collinsonana, varians, picicornis, nana, parvula, xanthura, convexiuscula, Afzeliella, Gwynana,* etc., and upon *Halictus æratus*, etc.

The other mode of parasitism destructive to the bees is where the parasite deposits its own egg upon the provender stored by the bee for the sustenance of its own young. The young of the parasite, either by being more speedily hatched or more rapacious than the larva of the sitos, starves the latter by consuming its food. This kind of parasites consists of several *Diptera*, but they are mostly bees which form a distinctive subsection of the family of true bees (*Apidæ*), the subsection being called the *Nudi-*

pedes or naked-legged, from their not having the necessary apparatus of hair upon the posterior thighs or shanks, for the conveyance of pollen wherewith to store their nests. Thus nature, having rendered them unable to perform this duty to their offspring, has imposed upon them the necessity of resorting to strangers to support them, and they are not led to it by idleness or indifference. These insects consist, with us, of six genera, the species of which are individually attached to some particular bee, who thus nurtures their young. They are, as a rule, gayer insects than those which they infest, and the genus most abundant in species is *Nomada*, which attaches itself chiefly to *Andrena*, although some of its species, especially the smaller ones, infest the species of *Halictus*, and one frequents *Eucera*. *Melecta* appears confined to *Anthophora; Epeolus* to *Colletes; Stelis* perhaps to *Osmia*, judging from the great similarity of habit; and *Cœlioxys* to the constructive *Megachile*. None of these parasites resemble their sitos, but *Nomada* is exceedingly different, being in its gay array more like a wasp than a bee. The only close approach in the appearance of a parasite to the insect upon which it is parasitical is in the resemblance between *Apathus* and *Bombus*, which are so alike that they were long continued to be united in the same genus, until the peculiar characteristic of the parasitical bees was detected, when they were readily separated. Although, cuckoo-bees as they are familiarly called, they could not be associated with the *Nudipedes*, because their posterior legs, though not pollen-conveying organs, are hairy; but the *Cenobites*, to which section they belong, have a peculiar and distinguishing structure of that limb. They are further separated from the *Nudipedes* by several frequenting the same nest, thus habitually associating with their sitos. Some of the *Chrysididæ* are likewise, as I shall have occasion to notice in the description of the habits of the genera, similarly parasitical upon some of the species of the family of bees. The genus *Mutilla* is also probably entirely parasitical upon bees, for *Mutilla Europæa* is a parasite upon *Bombus lapidarius,* from whose nests it has been dug in winter, by my friend the late Mr. Pickering, whose activity and accurate observation once promised to be very beneficial to the science, but he, like many others of my entomological friends, is now no more!

CHAPTER VI.

GENERAL PRINCIPLES OF SCIENTIFIC ARRANGEMENT.

The following rapid observations are addressed to those whom it is the desire that this series of volumes may induce to take up the study of Nature in a methodical manner. With this view, the merest summary of the principles upon which scientific arrangement is based, is here exhibited. The study requires method as a lodestar to guide through its intricacies, but it is one which, pursued simply as a recreation, yields both much amusement and gratifying instruction. It shows us that when we unclasp the book of nature, and wherever we may turn its leaves, every word, the syllables of which we strive to spell, is pregnant with the fruitfulness of wonderful wisdom, whose profound expression the human intellect is too limited thoroughly to comprehend.

Is there an arrangement that human skill could mend? Is there an organization that man can fully solve, or a combination that his mind can wholly compass? Do we not behold limitless perfection everywhere, but all so deeply mysterious. So exquisite are the feelings which the contemplation commands, that they imbue us deeply with the sense of the high privilege conferred upon the intellect by its being permitted to embrace a study, which, even pursued merely as a relaxation, inculcates in so serene and pleasing a manner such profound veneration and reverence.

To acquire the prospect of a possibility to unravel the exuberant profusion of the natural objects surrounding us, successive students of nature have endeavoured to systematize the seeming confusion in which her riches are spread about. Like has been brought to like, and gradation made to succeed gradation. Resemblances have been combined and disparities disjoined, until the labour of centuries has constructed of all the natural objects within the ken of man a vast and towering edifice, whose basis is seated at the lowest substructure of the earth which research has yet reached, but whose head ascends high into the empyrean.

All things have been collected, and arranged, and classed. Method has endeavoured to give them succession according to an assumed subordination. The labour of the great minds which framed the large theories of this vast branch of human knowledge, has permitted men of lesser powers of combination to abstract parts for special examination and investigation.

The study of natural science has progressively reached an extraordinary development, spreading in every direction its innumerable tentacula; to which the perfection of the telescope and of the microscope have still further added by the discovery of new worlds of wonder.

Just as language is systematized and made easier by grammar methodizing its co-ordinates and their relations, so natural science arranges its subjects into subdivisions of which genera and species are the lowest terms. The higher and more complicated arc of many denominations, which, notwithstanding, have for their chief purpose the simplification of the survey by assisting accurately to determine accurately natural objects individually. Once the clue of the labyrinth caught, the seeming intricacy of its involution vanishes; for when a clear conception of the general scheme is obtained, the solution of the parts is comparatively easy. The same principle rules throughout, however variously treated.

The large divisions of nature appear simple and distinct enough in their great frame, but when we approach their confines, close investigation discovers analogies and affinities, which, where the separation seems most apparent, create insuperable difficulties, and render linear succession, or distinct division, nearly an impossibility. *Here* we find parallelism, and *there* radiation, and elsewhere a complicated reticulation without subordination; and this is one of the great problems, which it is the office of the mature naturalist to endeavour to solve. The present work has to do, however, with but one small portion of the whole.

Thus we see that, in order to arrive at a knowledge of natural objects, a method must be pursued to avoid being overwhelmed by their multiplicity, whereby confusion would be produced in the mind which their methodical investigation tends to dissipate. Their abundance precludes the possibility of their being all equally well known, although it is very desirable to have a general, if even superficial acquaintance with them, that is to say, in the broad and distinguishing features of their large groups, for as to an accurate knowledge of all their species, it would be futile to attempt it. Possessing this general knowledge, the attention may be turned with greater advantage in any special direction, and that pursued to its entire acquisition.

Natural objects have been arranged in KINGDOMS,

ORDERS, CLASSES, FAMILIES, and GENERA, all deduced in their successive and collateral groups from characters exclusively derived from SPECIES; therefore to the accurate knowledge of *species* all endeavours must be directed, they comprising within themselves all the rest, although the characters upon which they themselves depend for separation from their congeners are the most trivial of any. Each combination, in its analytical descent, contains characters of wider compass than those which succeed it, and consequently embraces in that descent more species than the successive divisions; just as in the ascent, or synthetical method, the characters of every successive group gradually expand. Species being thus the only real objects in nature from which all knowledge springs, and in which exclusively all uses lie, other combinations being perhaps as merely imaginary as are the many lines which are drawn over the surface of the globes, it would imply that subdivisions merely lend aid to acquire more rapidly the details upon which they depend. We will, therefore, first turn our attention to species.

Both combination and subdivision are intended to facilitate identification, by aiding us to arrive at this knowledge of species; for each species represents a distinct idea, whose correct definition is important to the progress of accurate science. This alone permits observation to be attributed to its right object, and when properly recorded, the information is secured for ever from error or obscurity. It is not, however, the gift of every mind to discern accurately even specific differences, or to form skilfully generic combinations. The very best favoured by nature,—for it is a natural gift, although under high cultivation,—have sometimes a bias towards seeing more than actually exists. Hence varieties are often elevated into species, and species thus overwhelmingly multiplied; and genera are frequently framed upon vague distinctions.

Species are the basis of all natural science.

A species in zoology is a combination of creatures which unites the sexes, and these being two, the assumed existence of neuters in some instances does not invalidate this, it comprises two individuals having independent existence, but whose co-existence is indispensable to perpetuation, but which often, from their great differences, no single set of scientific characters will bind together, yet which must exist in some undiscovered peculiarity, that individuals may be able to distinguish their legitimate partners. The species, therefore, is a complete unit in its entirety, although consisting of two distinct beings, for in the large majority of cases in zoology these sexes are distinct, although their conjunction is, in the higher forms of life, indispensable for their continuance. In some of the lower forms of animal life they exist in union, and in the vegetable kingdom we perceive every possible combination and modification of this conjunction, and in both of these life may be perpetuated also by simpler processes.

The species may consist of any indefinite number of individuals, and no law has hitherto been discovered which regulates the relative proportions of the sexes, although it is very apparent that some recondite influence operates to control it. It is also extremely remarkable to observe how eccentric nature is in some species, and the extent to which she sometimes carries the variation of some particular specific type, and to which some species are singularly prone, and yet how rigidly in other cases she adheres to the particular specific form in the succession of generations, that even the shadow of a deviation from the typical distinction is scarcely to be discovered: a reason for this it is hard to surmise. We may, nevertheless, conclude it to be certain that true species are ever distinct, and can no more coalesce, however closely they may approach together, than can asymptotes.

Specific differences result from many characteristics,— from colour, clothing, size, and sometimes from peculiarities of structure; but these last are usually of a higher order, tending to indicate an aberration, slight though it be, from the normal generic character which holds the group together, thus implying a distinctive economy. This is sometimes called a subgeneric attribute, and there might be a reason, certainly, for not elevating such species to the full rank of genera, were genera equivalents, which they are not, and it merely remains an evasive admission of the doubt that attaches, except for the sake of convenience, to any subdivision, but the specific.

The species is thus the very last term of subdivision, the very elemental principle itself, which unites together as one, solely for the purposes of perpetuation, the two sexes of similar individuals, and without whose intercourse the kind or species would die out.

That some species greatly abound in individuals, as before observed, whilst others appear to be extremely limited, is an absolute fact, and not merely suggested by a defective observation of their occurrence, resulting from their rapid dispersion. It is verified by being noticed to occur where we know they would resort, as is exemplified in the case of some of the parasitical species of the insects herein treated of, and which are sometimes rare, even in the vicinity of the metropolis of their sitos, and where this also greatly abounds. In other cases, other species absolutely swarm where the similar attraction lies.

Even supposing species to be the sole natural division, we may accept the superior combinations as means to aid us to a gradually extending survey of the whole. Perhaps did we possess all the links of the vast chain of beings we should find genera, and every other superior combination, melt away through the intimate alliance of the succession of species that would obliterate the lines of separation, by making the sutures imperceptible; but what mind could compass the detail of such a limitless unbroken series? Their subdivision may therefore be accepted as a positive necessity, to enable us to compass their investigation. As it at present stands, with our imperfect knowledge of the entire series of species, these higher groups are indispensably requisite.

The specific diagnosis being the only sure basis upon which all our knowledge can rest, its accuracy is all-important, and requires a few observations. It comprises two parts—the specific *character*, and the specific *description*. The difference between these is, that the first is constructed with the extremest brevity consistent with its utility, is fluctuating and not permanent. The latter permits all the diffuseness needful to embrace a full description of the creature.

The object of the first is to establish the *present* identity of the species amongst all its known congeners—those associated in the same genus;—and that of the second to secure it in its perpetual identity, and segregate it from all future and contingent discoveries. The specific character admits, consequently, modifications to suit any extension of the genus, and in fact exacts it at the hands of all who describe new species. This many naturalists undertake without any apparent consciousness of the scientific responsibilities that attach to it, and whence results the confusion so much to be deplored, of the synonymy that prevails, constituting, as it does, such a Dædalian labyrinth. The describer of a new species is bound to cast around, and endeavour to know all that has been previously done upon the subject of the genus. He has to revise all the specific characters within the genus, and mould them to those he introduces, and he must insert these closest to their evident affinities. Thus, therefore, the describer's labour is not light, if to be of any value. The specific character, although thus varying, becomes a permanent utility, and only so fulfils its object,—that of rapidly showing, at a glance, the known species of a genus, and thereby permitting the speedy determination of the identity or distinctness of a compared object. If doubt should exist from this brevity, the specific description is at hand to solve it, by the amplitude and completeness of its details. Of course this mode of treatment is only suitable to monographs, or portions of the science discussed separately, and not to a general or universal survey.

The amount of toil thus saved to the describing naturalist, and to those who wish to name their specimen, the experienced only can estimate. This brevity of specific character is one of Linnæus's terse and valuable axioms, who limits its length to twelve words. The best examples, I think, that I can adduce in entomology, of valuable and exemplary specific descriptions, is Gyllenhal's 'Insecta Suecica,' which contains exclusively a description of Swedish Coleoptera; Gravenhorst's large monograph of European Ichneumons; Erichson's elaborate work upon the Staphylinidæ; and our own Kirby's 'Monographia Apum Angliæ.' Their perfection consists in fulfilling thoroughly all the above conditions, for if any doubt exist upon comparing your insect with their descriptions, you may be fully assured yours is not identical. The only drawback to the utility of Mr. Kirby's book is that he had to deal with insects variable in condition from many causes, and the variable state of the insect that may have to be compared; his description has evidently been made sometimes from a worn specimen, one that had been exposed to wind and weather, and sometimes from an insect in fine condition. Thus it is important that compared insects should be in an identical state to substantiate the comparison,—a difficulty which this family has specially to contend with, as these insects are more liable than almost any others to vary, owing to their specific character depending much upon pubescence, which is extremely subjected to many modifying influences, for the tinges and positive colour of the hair will much vary by exposure, as it is not possible always to capture a bright individual.

Taking specific description thus practically in its full and wide sense, it is requisite, for the purpose of avoiding repetition, that all the characters of the superior combinations should be eliminated, leaving it with those only which have not been thus absorbed, which now constitute its sole remaining distinctive specific peculiarities. Every species necessarily contains within itself, every character of every combination in direct line above it, although these have been gradually abstracted to form those several combinations which are arrived at successively in the synthetical ascent. Analytically, species are the last but combining element of all, although their most remote members. The whole system is an ingenious contrivance for breaking down a complex multiplicity of characters, to simplify the means of reaching all the collateral or adjacent species, that we may be able to determine identity or difference.

Entomology, and indeed natural history generally, uses three words, very much alike, but very different in signification and application. These are, *habit, habits,* and *habitat.* The habit is that peculiar character of identity, that *je ne sais quoi,* which marks all the species of a genus collectively, and which, in some cases, only the trained eye can detect. It is then seen instantaneously, and forcibly illustrates the extreme precision the study of the natural sciences tends to cultivate. Their utility, also, as a discipline to the mind, conjunctively with the keen accuracy which practice gives the sight, are qualifications not lightly to be esteemed.

It is from such absolute control of detail that the most efficient power of generalizing emanates, which, when it has once become habitual, gives, from its rapidity, an almost instinctive facility, as its inevitable concomitant, for both synthetical and analytical survey. The mind thus becomes strengthened by vigorous exercise, and has always, for every purpose, a powerful instrument at command, often used unconsciously, but always effectively. Thus is *habit,* once correctly perceived, ever retained.

The *habits* are the peculiar manners and economy of a species; and the *habitat* is the kind of locality the creatures affect, such as hill or plain, wood or meadow, forest or fell, hedgebank or decaying timber, sand or chalk or clay, and ground vertical or horizontal; and the *metropolis* of a species—another term in use—is the centralization of the general habitat where the insect either nidificates collectively with its fellows, or, where, from any other cause, it may be found in its season, usually in profusion. But good fortune does not always attend the discovery of this locality.

It is by the acquired skill of perceiving habit, that a large and confused collection may be sorted rapidly, or fresh captures immediately placed with their congeners, without the necessity of going tediously through all the descriptive characteristics. Incidental errors are afterwards speedily corrected. It is then that the specific character exhibits its utility by enabling us at once to distinguish the new from the old.

The concentration and summary of the specific character is the name of the species, or trivial name as it is sometimes called, which is, as it were, the baptismal designation that attaches to it always afterwards, and is contemporaneous with the introduction of the creature into the series of recognized beings.

Upon the revival of the study of natural history, when learning dawned after the night of the Middle Ages, much difficulty attached to the imposition of discriminative names. The works of the ancients were ransacked, and endeavours made to verify and apply the names they had used. Ray published a vocabulary of such names. But the ancients never studied natural history in the systematic way pursued by the moderns; they did not want the skill, but they wanted the facilities. Anatomy and physiology had not made the progress necessary to aid them in the pursuit, and the assistance all these sciences obtain from optical instruments was barred from them. The names they gave to natural objects were vernacular names, which, like our own vernacular names, applied rather to groups than to species, and have in consequence ultimately become the names of genera. But this was the work of time, with which discovery progressed. As these discoveries were made by the new cultivators of natural history, they added them to those which they resembled, by some brief distinctive character adapted to the momentary exigency, such as *major*, or *minor*, etc.; and these additions were constantly treated as varieties of the species, whose name headed the list by the designation first adopted. Discoveries still continued, which were compulsively arranged with the predecessors they most nearly resembled, until resemblances vanished, and the boundaries fixed by the assumed correct application of the names thus derived from the ancients were passed, and there was an overflow on all sides.

To meet this difficulty, the new discriminative name had to be moulded into a phrase to correct its exceptive peculiarities, and specific names became descriptive phrases, the bulk of which no memory could retain, and which usually were neither clear nor expressive. Thus genera were continually treated as species, and species as numbered varieties, with long distinguishing descriptive phrases.

So it remained till day dawned, and the great luminary of systematic natural history rose with a bound to irradiate the obscurity of science with his subtile and vivifying beams.

This was LINNÆUS, to whom we owe the binomial system, wherein, by means of two words only (the generic or surname, and the specific or baptismal name), the recognition of a species is perpetuated; for Linnœus truly says, "*Nomina si nescis, perit et cognitio rerum.*"

By a law tacitly admitted, but universally recognized, for the sake of securing to a name its intangibility, no two genera in the same kingdom of nature may be named alike. There is, therefore, if this rule be observed, no fear of similar names coming into collision in the same province, and thus producing confusion. A ready means to prevent the possibility of such mischance is the admirable work which has been published by Agassiz, with the assistance of very able coadjutors, in the 'Nomenclator Zoologicus,' which is a list of all the generic names extant in zoology, exhibiting what names are already in use either appropriately or synonymously in this great branch of the natural world, and if this work receive periodically its necessary supplements and additions, no excuse will remain for the repetition of a name already applied. The most defective character in this laborious work, is the frequent incorrectness of its etymology of the names of genera. It would be, perhaps, without such aid, too great a labour to require of the describing naturalist, or it might not be otherwise even practicable for him, to ascertain whether the generic name he purposes to impose be, or not, anticipated. The penalty of its being superseded is understood to attach to the imposition of such a name, for the alteration may be made with impunity, and thereby it becomes degraded to the rank of a mere synonym.

Nomenclature has thus, by the happy invention of Linnæus, been made a matter of the greatest simplicity, conciseness, and lucidity, and to him, therefore, our gratitude is due.

An indispensable branch of nomenclature is *Synonymy*,

which, briefly, is the chronological list of the several names under which species or genera may have been known. This diversity of names has originated in several ways,—from indolence, or ignorance, or excessive refinement. The views of systematists will differ in the collocation of creatures; hence, sometimes what had been previously divided will be recombined, or divisions into further groups be made of what had been before united. Both processes will necessarily produce synonyms; the recombination of what had been separated reduces the names of such groups to the rank of synonyms of the old one from which they have been disjoined. In the latter case the old name will be retained to the typical species merely, and be also made a partial synonym of the names of the new generic groups: or, indeed, it may happen that the same creature has been described generically, unknowingly, by two different persons, about the same time. By another recognized rule in nomenclature, the 'law of priority,' the name given by the first describer is accepted, and the other consequently falls to the condition of a synonym.

With respect to specific synonymy, many causes conduce to it; namely, an imperfect description which cannot be clearly recognized, reducing it to that category, with a mark of interrogation appended; subsequent description when want of tact has not discerned the identity of the old one; indolence in looking about for works upon the same subject; inability to obtain access to books wherein they may be described, owing either to their costliness or to their obscurity, or by lying buried in some collapsed journal, or the poverty of our public libraries, etc. etc. But however thus lost sight of, or wilfully ignored, the name still retains vital elasticity, for the describer has not thereby lost his rights, but revives to them with all due justice upon the cessation of this coma. The really culpable among such describers are those who neglect to look around them to ascertain what has been done, and this course is sometimes illicitly adopted to obtain a fleeting and meretricious fame, by the description of ostensibly new species, which critical investigators soon detect to have been long since known and very ably described.

Thus, a complete synonymy, which can almost only come within the province of a monograph, would give, chronologically, the entire history of a species under all the names it has been known by in the several works in which it has been published. Nature is so uniform and stable that Aristotle's descriptions can be clearly recognized, therefore there is no fear that whatever may have been synonymously, but yet correctly recorded of the economy of a species, can possibly be lost when once registered in the archives of science.

The working out of a correct synonymy is an ungrateful task of much labour, for few appreciate it, and not many use it, although when thoroughly elaborated it is so extremely valuable.

A further rule in nomenclature is, that the generic name must always be a substantive; and it is always desirable that the specific name should be an adjective. In the event of the imposition of a proper name, which is sometimes done to record a private friendship, but improperly so, for it is a distinction due only to promoters of the science, the genitive form must be adopted.

The next grade in ascent from the species is invariably the Genus, for subgenera, like varieties in species, are not uniformly present, but are mere contingencies, even if they do properly exist.

Why some genera abound in species and others are so limited is as difficult to determine as the differing numerical abundance of individuals in species. That long genera (genera numerous in species) may be the result of natural selection, as Mr. Darwin surmises, and the offspring of a common parentage, is contradicted, not merely by peculiar although sometimes slight dissimilarities of habit, combined with size and colour, but also if any lines of demarcation are to be admitted, it is possible, were their generic similitude to be subjected to severe test, they might present characteristics normally discrepant and suggestive of further division, although the habit may be very like.

The generic grouping is effected by structural peculiarities, which are essentially of a higher class than the characters of specific separation, these being determined by colour, pubescence, sculpture, etc. etc.; specific characters combining only individuals with such peculiar inferior resemblances. The generic characters thus establish groups of species allied only by such more general character and similarity, but conjunctively of one permanent habit, although the members of the genus may differ somewhat in habits, and so on of the higher groups into which insects are collected, each group in its ascent upwards presenting characteristics of a wider range than those of the descending series. And so, by degrees, we rise until we reach the characters which combine the whole order. The process is necessarily and imperatively synthetical, for the whole foundation is based upon species, and thence emanates the supposition that only species exist.

The *type* of a genus is that species upon the characters of which the genus was originally framed and named, and theoretically, however generic groups may be subsequently divided to suit views or to meet systems, the primitive generic type is assumed to retain the primitive generic name. It is much to be doubted whether, in every case, the type is the true pattern, or leader, or centre of the group called the genus; nor is it likely if genera be natural groups. It has usually been accident which has dropped upon the favoured species, and not a well-calculated and thoroughly digested selection, and which, although accepted, will require emendation or change if the whole collective series should ever be obtained.

It is the necessary result of the imperfection of our intellect, and one of the dominant conditions of overruling time, that one thing must follow the other. It is, therefore, neither an expressed nor even an implied inferiority that puts one species before the other in a generic group; or one genus before the other in their successive order. Affinities may lead both species and genera in varying directions, although treated descriptively as of linear succession, in which order they are usually arranged, but this is unavoidable and therefore not derogatory. It is for the mind to conceive their radiation from a type, or their parallelism with other forms, even in the connection of affinity, and not merely of analogy, for the latter can be expressed even in arrangement.

Thus encouragement attends the beginner at the very outset of his study, and the prospect of a wide field for discoveries, in all directions, lies open to him.

The FAMILY, after the GENUS, is the next natural group at which we arrive, proceeding synthetically. Its characters, succeeding to those of the ORDER, group together collectively the largest numbers of forms that in their several combinations are the most nearly equivalents, and may be almost paralleled in that quality to the alliance of species. Ascending from species, the naturalist scarcely hopes to find in the groups formed above them strict parallelism, although, to be logical, it should be so, and, where the combinations are most natural, it is most nearly so. Thus we do not again distinctly reach equivalents until we arrive at these families, which from linking together associations usually combined by an identity of instinct and functions, attach to themselves greater interest, and form alliances pointed out by the finger of nature itself, which are therefore exempted from the arbitrary caprice of the constructive systematist.

It does not follow that families should be even nearly numerically equivalent, for a family may contain a few or a multitude of genera and species, or a multitude of genera and few species, or also a multitude of species and few genera. Families comprise groups of forms to which nature delegates the execution of certain duties and offices, and whether specifically numerous or few, we may assume they are sufficient for the object intended. If we can reach the motive that controls the peculiarities of the group, it is a golden key to the explanation of the structure of its constituents, and, perhaps might furnish us, if not with a positive clue, yet with a surmise as to the functions of the collateral groups of which it forms a member, and which diligent observation may accurately determine.

Families, to be natural divisions, should stand in the same relationship to genera as species do, but from the opposite side, whatever the subdivisions are into which they may be separated, for the sake of convenience, and as descending grades whereby to arrive with greater facility at their genera, just as the species of the latter are also sometimes grouped, that they may be reached with greater ease. These subdivisions of families have no analogy with the varieties which species occasionally throw off, although they may be as irregular in their occurrence; that is to say, in the association of a group of families arranged in their series of most proximate affinities, the first may present subdivisions, others, in irregular occurrence, may not require them,—just as in the species of a genus, arranged also in the series of their closest resemblances, one will present a stringent adherence to the specific type, or all may do so, or all or some may have a tendency to vary. Groupings of species are, however, of a less natural character usually than are those of families, and generally are artificial, being capriciously made to break down long genera, that the required species may be more readily arrived at.

The characters which group families differ *inter se*. Thus in the Order *Hymenoptera*, the family of the bees is essentially framed upon their most distinguishing peculiarity—the tongue,—which in other families becomes of secondary importance. In some the neuration of the wings, their mode of folding, the form of the eyes, conjunctively with other peculiarities of general structure, etc. etc., which point to the differences in the economy that accompany all these, have successively the same prominent position which the trophi take in the family of the bees.

I have already recently alluded to the relations of affinity and analogy, and it is desirable that some notion of the meaning and bearing of these terms should be given, as, in the majority of modern works on natural history, use is frequently made of them.

On carefully surveying any class or order of creatures, the mind speedily becomes impressed by observing certain similitudes out of the direct line of continuous connection, and therefore remote from the strongest connecting links of positive relationship in the methodical series. Induced thence to inspect them more closely, we presently ascertain that what we at first conceived might be an error in their collocation, arises from very strong resemblances in certain particular features, but which are less important than those which directly unite them, and may not be permitted to interrupt the order established. It is, however, equally evident that they indicate relations which may not be neglected.

Thus, although the succession be direct in the evolution of its primary characteristics, the prominent features which so present themselves establish the conviction of the existence of connections oblique to the straight line, but all embraced within the normal conditions which bind the group together. These are called relations of affinity. Pursuing them, it is sometimes observed that nature, as it were, returns upon itself, reproducing similar notes in another key.

These indications have led philosophical naturalists to surmise that the true arrangement of natural objects is in groups, and not in a straight and continuous line.

Several schemes have been suggested for the purpose of giving uniformity to these groups, making them equivalents by associating together the same numbers of allied forms, which again return in a circular series upon themselves, and impinge upon other circles at the parallel points of their circumference by affinities less direct than those which unite them within their own circle.

Many novel views and interesting combinations have been thus elicited, showing that very strong affinities lie in very divergent directions, but no system has been hitherto devised which overrules the conflicting difficulties that attend these arrangements. Whatever number may have been adopted to bring nature within this circular system, it has always been found that some, or several members, both in the circles themselves, or in their series, is as yet deficient, and awaits either discovery or creation.

The pursuit of such views stimulates profound investigation, and may lead to valuable discoveries that will eventually give a loftier and more philosophical character to the study of natural history than it has hitherto possessed, and make it an attraction to the highest class of mental powers. The key to the universe hangs at the girdle of the veiled goddess; and happy the student who shall achieve possession of it, and unlock the mysteries to the reverential gaze of mankind.

The *relation of analogy* is different in *kind*, although the general affinities which bind a class together are necessarily affinities in the widest construction of the term; but the class being resolved into its elements, those affinities, thus dissevered, no longer retain the uniting links whereby the mass coheres. They, more correctly, stream from their origin in parallelisms rather than in a continuous and uninterrupted current; and these parallelisms present resemblances often of a merely superficial character. As strong an instance as I can adduce is possibly the analogical parallelism of the *Pentamera* and the *Heteromera* in the *Coleoptera*, which are, however, bound by the common affinity of being all beetles.

It is, nevertheless, often difficult to determine between the relationships of affinity and analogy, for groups even in close contiguity may also possess both. Thus, the normal *Ichneumones* have their analogues in the *Ichneumones adsciti*, if the comparison be restricted to themselves, but these revert into the relationship of affinity when a comparison is instituted between them and the adjacent groups on the one side of the *Tenthredines*, or on the other of the *Aculeata*, with which, when a relationship presents itself, it is merely one of analogy. So, also, within the pentamerous *Coleoptera* we have a relationship of analogy between the *Staphylinidæ* and the *Histeridæ*, but it becomes one of affinity when it unites them within this section of the class.

Innumerable other instances might be given readily, but these will suffice to convey a notion of the relative meanings of the terms, 'relation of affinity' and 'relation of analogy,' which is all here aimed at.

The problem naturalists have to solve is, "What is the natural system?" We can clearly see that the systems adopted are not Nature's, that they are essentially imperfect, and that the science, even with all the force of the intelligence that has been applied to it, is far from having attained perfection. It still awaits the master mind that shall cope with its difficulties, determine its intricacies, and, threading the labyrinth, guide his enthusiastic disciples into the adytum of the temple.

The subjects here brought under view admit of very considerable development, and of strictly didactic and methodical treatment. It has been my object only to gossip upon them, that I might stimulate curiosity to undertake systematic study, by showing how interesting it may become if earnestly pursued, being so fraught with instruction of large compass.

Works on natural history have divers objects in view, and may be intended either for popular and general distribution, or for special scientific purposes, and in each case the mode of treatment will materially differ. Many purposes may also be intended to be severally met in the strictly and rigidly scientific treatment. They may be either general methodical arrangements treated superficially, having no other design than to give a sort of bird's-eye view of the subject in its wider distributions and broader landmarks, or they may treat of portions of the large subject more specially; again, they may constitute monographs of varying extent from a family to a genus; or they may comprise loose descriptions of new species of old and well-established genera; and some such, conjunctively with new species, establish likewise new genera, indicating, at the same time, their proximate position in the general series. The two latter classes are usually the appendages to voyages and travels in distant unexplored countries, or are the result of a careful collection of neglected tribes at home. Each, thus, with its special application has its special construction; but in the case of new species, I would strenuously counsel a full and complete description, and urge as imperative the construction of a specific character, formally framed to meet the condition of the science, based upon the precise antecedents and existing state of the genus to which such species belong.

Even assuming that the knowledge of species is the essential foundation of the science, the preceding observations show that there is a higher knowledge connected with the pursuit than this mere knowledge of species, and yet from which it emanates. There is a higher object to be achieved than the accumulation of a store of them, arranged in seemly order, set with manifest taste, and named in accordance with the accepted nomenclature. These are extremely pleasing to the eye, but the intellect languishes over them in unsatisfied desire, craving more solid aliment. There is besides room for observation on every side, either confirmatory or original, and both are much needed, and must be considerably augmented before it is accumulated in satisfactory abundance; and until this be procured, existing systems can be viewed merely as temporarily useful, for until all that nature can teach shall be exhausted, perfection cannot be attained.

The many kinds of knowledge which the study subserves, and the recreation and pleasure each affords, are a sufficient reply to the sneering *Cui bono?* of its detractors, who, when they urge that it occupies time which might be more profitably employed, present themselves but as the priests of the Fetish of the age, and may be told that we use it only as a relaxation to necessary worldly toils. When pursued, in cases where it can be so, in unmolested security, is there a more salutary pursuit than that which inculcates the high veneration and love which the study of nature should inspire towards the Great Parent of all? What can compete with it in other studies? The investigation of the works of the Almighty lead directly to the steps of the altar of religion, and there we find the study of the Works confirmed by the precepts of the Word, both inculcating humble reverence and fervent love. Thus pursued, is it not a reply to every cavil?

CHAPTER VII.

BRIEF NOTICE OF THE SCIENTIFIC CULTIVATION OF BRITISH BEES.

WITH the great JOHN RAY dawns the scientific cultivation of British bees. Before his time, the only entomological work which had been published in England was Dr. Mouffett's 'Theatrum Insectorum.' In this work there is an ample account of the domestic bee, with gleanings from many sources of some of its habits and economy, but there is no notice of any insects, excepting some species of the genus *Bombus*, which may be at all consorted with the social bee by affinities of structure or identity of function.

In Ray's correspondence with his disciples and friends, we have straggling observations upon the habits of a few wild-bees, especially some jotted down by his diligent pupil, the distinguished Francis Willughby. It is in Ray's posthumous 'Historia Insectorum,' published in 1710, at the instance of the Royal Society, that we first find collected together all that had been previously known of 'British Bees.' In that work he describes them systematically. He there arranges the bees into *Apis* and *Bombylius*, which may be regarded almost as genera.

He divides *Apis* into what may be considered as two sections, *Apis domestica* forming the first, and the second containing his *Apes silvestres*, or wild bees. Nine of these are described and numbered consecutively, which are followed by eleven descriptions unnumbered, some of the latter having been supplied to him by Francis Willughby, whose initials are attached to these, and amongst which we find the description of the willow bee, subsequently, from this cause, named by Kirby, from its original describer, and now universally known as *Megachile Willughbiella*.

Ray's second genus is *Bombylius*, identical, as far as it goes, with the modern genus *Bombus*, excepting that it includes an *Anthophora*. He here describes nineteen, all numbered. Ray's names are phrases, the mode of describing then prevalent in all the natural sciences, until the happy introduction of the binomial system by the great genius of natural history—LINNÆUS. These phrases are almost tantamount to the modern specific character; but Ray unfortunately attaches no size, yet size might have lent some aid to their modern determination.

Mr. Kirby was able to identify and introduce into his synonymy only a few of Ray's insects, from the defectiveness of the descriptions; the following embrace all that could be verified:—

No. 1 of the *Apes silvestres* is our *Anthidium manicatum*; No. 3, the male of *Anthophora retusa*, the female of which being No. 4 of his *Bombylii*; No. 4 of the *Apes* is *Andrena nitida*: these comprise all of those numbered which could be recognized. The first of the unnumbered is the male of *Eucera longicornis*; the fourth is *Melecta punctata*; the sixth is *Colletes fodiens*;

the seventh is the male of *Osmia bicornis;* and the ninth the celebrated *Megachile Willughbiella.*

In *Bombylius* No. 1 is *Bombus lapidarius;* No. 2, *B. Raiellus,* named by Mr. Kirby in honour of its great describer; No. 3 is *B. muscorum;* No. 4 is the female of *Anthophora retusa,* as noticed above; No. 5 is *Bombus terrestris,* as is also No. 6; No. 7 is the male of *B. lapidarius;* No. 8 is *B. pratorum;* No. 9 is *B. sylvarum;* No. 10 is *B. subinterruptus;* No. 11 is *B. hortorum;* No. 13 is *B. Francillonellus,* and No. 17 is *Apathus Barbutellus.* Thus ten of the *Apes silvestres,* and six of the *Bombylii* are unidentified, and those recognized may be placed correctly, by the aid I give in attaching Mr. Kirby's synonymy to the list of species added to each genus below.

Nothing of any moment thence intervened, until the Rev. W. Kirby, of Barham, in Suffolk, made a careful and earnest collection of the 'British Bees,' with a view to their scientific description and distribution. Stragglers were to be found in many entomological cabinets, and some of their habits had been observed and recorded by patient and attentive naturalists; but these collections were small, very imperfect, and widely dispersed, until Mr. Kirby's energy and activity nurtured the idea, and carried it into execution, of bringing into one focus the scattered notices and vagrant specimens he had seen about.

The diligence he himself exercised in procuring all the individuals he possibly could, by continued collecting during a succession of years, enabled him, in the course of time, to add considerably to those he was already acquainted with, either in collections, or through dispersed notices. The growing bulk of his store suggested his looking around for guides to their methodical arrangement, as a clue to what might have been observed of their habits. Finding no such assistance, and nothing to meet his wants, for Linnæus's notices were too few, and Fabricius's labours too inconsequential, he determined to aid himself by elaborating their distribution upon the basis of the principles established by Fabricius himself, but which this celebrated entomologist had worked out so inconclusively as to make his system an indigested mass heaped together in the greatest disorder.

Mr. Kirby's patience and diligence, although working only upon the same principle, speedily brought into lucidity and order the obscurity and confusion that had prevailed. By one of those strange coincidences which have been remarkably recurrent in scientific invention and discovery, Latreille, in France, was at the same time arranging all the bees known to him, by a process precisely similar to that adopted by Mr. Kirby. He consequently arrived at exactly the same results, with this difference only, that what Mr. Kirby calls genera are to Latreille sub-families, and the sections which Mr. Kirby was induced to form in his genera, from their structural differences, and which sections he called families, inconveniently indicating them merely by letters, asterisks, and numbers, were formed by Latreille into genera, and to which the latter either applied or adopted names, or framed new ones, when deficient; these however are essentially genera, with all their discriminative characteristics, for they bring together the very same species in both cases. This clearly exhibits the beauty and certainty of the principle upon which each had worked out his distribution, both being based chiefly upon the structure of the trophi, or the organs of the

L

mouth, but which Fabricius, its projector, had, singularly enough, failed to accomplish successfully.

Both works were published in the same year, 1802 (An X. of Latreille's book), unknown to each other, but Mr. Kirby's sprang into life in matured perfection, like the imago of the bee itself, whereas Latreille's labours were progressively nursed to maturity in successive publications, until they received their final elaboration in 1809, in the fourth volume of his 'Genera Crustaceorum et Insectorum,' whose successive stages were, first, the notice appended at the end of his 'Histoire des Fourmis' in Paris in 1801, and then in the thirteenth volume of his 'Histoire Naturelle des Insectes,' in 1805, a supplement to Sonnini's edition of Buffon, and then in the 'Nouveau Dictionnaire d'Histoire Naturelle.' Even thus the subject was not so amply discussed, although applied more extensively, and made to embrace all the bees, exotic as well as European, at that time known, as it had been done in Mr. Kirby's model work, which leaves nothing to be desired but the naming of his anonymous subdivisions, and a little more artistical skill in the execution of his plates. The terminology used by him also differs from that subsequently adopted through foreign influences, but which is readily reduced to his standard.

The merits of the work greatly transcend these trivial deficiencies, for it is a "canon" as invaluable to the entomologist as the celebrated canon of Polycletus was, and the Phidian marbles still are to sculptors. Of course observation has greatly reduced the number of his species by their due association with legitimate partners, which, from their dissimilarity, he was compelled to separate, as only successive observation could prove their identity.

More extensive collecting has also shown that some of his species are merely varieties of others, which have thus been brought to their authentic type. This also could only be proved by experience, for it is remarkable how very Protean some species are, whilst others are almost rigidly unchangeable. Evidently there does exist a line of demarcation between distinct species, which only requires to be diligently sought to be found, obscure as it may appear to be, but which the insects themselves obey, for however closely species may sometimes approximate, yet I do not believe, as I have before expressed, that they ever permanently coalesce, and that they are always as distinctly separate as are asymptotes.

As Mr. Kirby's work is in few hands, or perhaps not readily accessible, I will give here a summary outline of it, with the names of the genera with which his families coincide.

In this work he established only two named genera— *Melitta* and *Apis*.

His genus *Melitta*, which is equivalent to the subsequent subfamily *Andrenidæ*, he divides into two sections, * and * *, the first containing two families, *a* and *b*, (these we call genera, and they are now named *Colletes* and *Prosopis*); the second section * * contains three families, *a, b, c*, (*a*, is *Sphecodes*, *b*, *Halictus*, and *c* comprises our three genera, *Andrena*, *Cilissa*, and *Dasypoda*.)

His genus *Apis* he also divides into two sections, * and * *; the first is subdivided into two families, *a* and *b* (our genera *Panurgus* and *Nomada*); and the second is divided into five subsections, *a, b, c, d, e*; *a* and *b* constitute families (our genera *Melecta* and *Epeolus*). The subsection *c* is divided into two parts, 1 and 2, the first containing the two divisions *a* and *β*, each comprising a family (our genera *Cælioxys* and *Stelis*); and the second is divided into the four families, *a*, *β*, *γ*, *δ*, (*a* being the modern *Megachile*; *β*, *Anthidium*; *γ*, *Chelostoma* and *Heriades* conjunctively, and *δ* is our *Osmia*). The subsection *d* has two subdivisions, 1 and 2, the first being a family (our *Eucera*); and the second is divided into the two families *a* and *β* (*a* comprising our *Saropoda*, *Anthophora*, and *Ceratina*), and the family *β*, consisting of the genus *Xylocopa*, then supposed to be indigenous, but whose native occurrence has not been substantiated.

The fifth subsection, *e*, is split into two divisions, 1 and 2, each containing a family (1 is our *Apis*, and 2, our *Bombus*).

In this last of his families Mr. Kirby had already noticed, with the same sagacity with which he had previously conjectured the cuckoo-like habits of some of the solitary bees, the distinctive structure of some of the species, which incapacitated them from providing the sustenance of their own young, and which thus reduced them to the same category; but he left the idea in its supposititious condition, being too modest to use it as a mark of separation, but which Newman, on our side of the Channel, and St. Fargeau on the other side, subsequently, and both nearly about the same time, but with the advantage in favour of Newman, distinguished, and separated generically, respectively by the names of *Apathus* and *Psithyrus*; the former, having the priority, is adopted, according to the rights of precedence in nomenclature.

The above description of Mr. Kirby's system will perhaps be difficult to understand, unless I append the naked scheme itself, which is as follows:—

SCIENTIFIC CULTIVATION OF BRITISH BEES.

MELITTA.

* { Family *a*.
    „ *b*.

** { „ *a*.
    „ *b*.
    „ *c*.

APIS.

* { Family *a*.
    „ *b*.

** {
  Subsection *a*. Family *a*.
    „ *b*.   „ *b*.
    „ *c*. {
       1 { Family *a*.
             „ β.
       2 { „ α.
           „ β.
           „ γ.
           „ δ.
    }
    „ *d*. {
       1 Family
           „ *a*.
       2 { „ β.
    }
    „ *e*. { Family 1.
             „ 2.
}

Mr. Kirby could scarcely have considered that there were more than two series of equivalents in this scheme, the first being the great division into the two genera; and the second, the final division, where his analysis terminated in his families, which, with some further slight subdivision, as shown above, constitute our present genera. The synthetical combinations which the arrangement presents, as we ascend from his families, result from an almost arbitrary selection of characters and certainly are not equivalents. The whole method is very perplexing; for, to cite an insect for the purpose of making a communication, it would have to be preceded by its whole array of subdivisions. Thus *Megachile Willughbiella*, which is now so compendiously noticed by the binomial system, would have to be quoted as *Apis* * * c, 2, a, *Willughbiella*, and so with the rest.

Although I have strongly applauded the 'Monographia Apum Angliæ,' as an excellent treatise wherever I have had an opportunity, the praise is to be applied to the correct care with which both the family descriptions and the specific descriptions are elaborated; whilst Mr. Kirby's timidity in fearing to depart from the course of his masters, Linnæus and Fabricius, by establishing a multitude of genera unrecognized by their authority, although every one of his families is pertinently a well-constituted genus, is much to be deplored. He has thus lost the fame of naming the offspring, of which, although legitimately the parent, he was not the sponsor. But he has won the higher renown, as I have elsewhere remarked, of his work being a canon of entomological perfection.

Notwithstanding that this very elaborate, and, to some extent, artificial method is based upon a plurality of characters, and apparently upon such as most readily presented themselves to substantiate the feasibility of subdivision indicated by habit, it is very remarkable in having brought the series into more satisfactory sequence than that presented by Latreille and his modifiers. *Panurgus* here holds its permanent post as the connecting link between the *Apidæ* and *Andrenidæ*, pointed out by nature in its close resemblance to *Dasypoda*. But this genus, however, establishes for itself a stronger affinity to the *Apidæ*, exclusively of that presented by the folding of the tongue in repose, in its presenting immediately the large development of the labial palpi which is peculiarly characteristic of this subfamily.

All the cuckoo-bees then follow in order; these are succeeded by the true *Dasygasters*; after which come Latreille's *Scopulipedes*; and the series is wound up by *Apis* and *Bombus*.

Mr. Kirby, I suppose, was induced to associate in the same section *Panurgus* and *Nomada*, from their resemblance in general habit, which in both conforms to the type predominant in the *Andrenidæ*, although they are thence dislocated by the differences in the important organs of the mouth, which verify in this case the seeming paradox of a part being greater than the whole; for these are certainly of greater relative importance to the economy of the creature than mere general habit, and to which all the peculiarities of structure finally converge, for the purpose of giving it what it thence acquires, its own proper and distinctive place in the series of created beings.

The most extensive work since published upon bees generally, is that treating of the *Hymenoptera* universally, written by Le Pelletier de St. Fargeau, and comprised in four thick octavo volumes, contained in the 'Suites à Buffon.' In this work both the genera and species of our bees occur, of course conjunctively with the rest, but its utility, especially to the beginner, is materially diminished by the peculiar systematic views of the author. The distribution of the Order is framed chiefly upon the economy of the insects, which is not so tangible as structure, and blends very heterogeneous forms,—widely separating, in some cases, structural affinities, and sometimes uniting discordant habits. Wasps and bees we here find intermingled, and to commence study with this work would much perplex the student. It can be used beneficially only when some progress has been made in the pursuit.

The only British entomologists who have treated of the bees since the time of Mr. Kirby, are Stephens, Curtis, Westwood, and Smith,—the first in his elaborate 'Catalogue of British Insects,' published in 1829; and the second in his 'Guide to the Arrangement of British Insects,' published in 1837. The arrangement of the family of bees in both these works is exceedingly arbitrary and without any obvious reason, either as regards the consecutive order of the genera or species. This originated possibly in their personal rivalry, which led them to make their systems as dissimilar as they could, and as unlike the true order as they could well dispose them. Both arrangements are certainly far beneath criticism.

In the Synopsis of Westwood, at the end of his 'Guide to the Classification of Insects,' published in 1840, and in Smith's 'Catalogue of the British Bees, contained in the Collections of the British Museum,' published in 1855, we have Latreille's distribution, with slight modifications, to which I shall not advert at present, but which I shall discuss in my next chapter, where I shall introduce the arrangement I myself propose for the combination of the genera of British bees.

CHAPTER VIII.

A NEW ARRANGEMENT OF BRITISH BEES, WITH ITS RATIONALE, AND AN INTRODUCTION TO THE FAMILY, SUBFAMILIES, SECTIONS, AND SUBSECTIONS.

If perfection of instinct, and an organization exquisitely moulded to a complete adaptation to the many delicate and varied functions of that instinct, as well as to the exercise of every faculty incidental to the class, be certainly a proof of pre-eminence, we may justly claim this position for the Order Hymenoptera. There is no characteristic in which they are deficient, nor any in which some of the members of the Order do not transcend in aptitude the insects of all the others.

If they have not been placed at the head of the class Insecta, it has been because systematic convenience did not permit the transposition, on account of the interruption it would have caused to the convenient linking of the rest in a consecutive arrangement. Yet are they the most volatile fliers, the most agile runners, the most skilful burrowers, and consummate architects.

The beauty resulting from the combinations of symmetry of form, elegance of motion, brilliancy of colour, and vivacity of expression, is to be found exclusively amongst them. Either in the velocity of their flight, or in its playful evolutions and graceful undulations, they are unsurpassed, and they hover in the execution of their designs with pertinacious perseverance. No insect structure can more thoroughly exemplify the most appropriate adaptation to its uses, and the most admirable elegance in the formation of the means of execution.

I thus claim for them, and which I think I may without infraction of dispute, the distinctive rank amongst insects.

Having fixed the station of the Hymenoptera generally, we have next to seek the relative rank of the natural divisions into which they readily separate.

Taking structure and instinct conjunctively, there can be no doubt that the first position will be conceded to that division of the Order which comprises the aculeated tribes—those armed with stings,—some of whose members, in each of the three large divisions into which they fall, being social, that is, living in communities, organized by a peculiar polity or administration.

These aculeates divide into, first, the *fossorial Hymenoptera,* or burrowers; and the equivalent branch the *Diploptera,* or wasps, distinguished and named from their folding the superior wings longitudinally in repose; secondly, the *heterogeneous Hymenoptera,* or ants, named from the dissimilarity either in size or structure of their females, a peculiarity incidental to all the social Hymenoptera, but living in community is more peculiarly characteristic of this division, it being in the other divisions restricted to a few genera only, whereas here the solitary habit is the exceptional. In all cases of socialism there are three classes of individuals,—males, females, and abortive females. In the other social kinds of

Hymenoptera, these abortive females, called neuters, perform the labours of the community, and they are always winged; whereas amongst the ants they are never winged, and they constitute civil and military departments, the former attending to domestic matters, and the latter making predatory excursions to enslave the inhabitants of other communities, to aid their civilians in their many duties.

The third and last division of the aculeate Hymenoptera contains the *Mellicolligerae*, the bees, or honey-gatherers.

Thus each division of the aculeated Hymenoptera is closely linked to the others by the strong affinity of the social habits of some of the genera of their several families.

The food of these three divisions of the aculeated Hymenoptera differs considerably, the Fossores being raptorial flesh-feeders, which hunt down and destroy their prey, and supply it as food to their young; the *Heterogynae* are omnivorous,—grain, fruits, or carrion being equally welcome to them; but in these climates I am not aware that they destroy life, although their wide migrations within the tropics are undertaken in the very spirit of the Huns and Vandals, for they devastate everything they come across; but the whole family of bees are exclusively honey-feeders without any carnivorous propensities, and use their stings merely as weapons of defence.

Although all the social aculeates are edifiers, and although the wasp in its *papier mâché* domicile may vie with the honey-bee in capacity and skill in the structure of the hexagons of the habitation it erects or suspends, which are as perfect, and almost as delicate, although fabricated of a coarser material than those within the hive, and wherein also the several compartments form a more homogeneous unity, and the uniformity of the several layers or floors is more in accordance with architectural symmetry,—yet must the palm of precedence be accorded to the bee, from the more elaborate and perfect development of the social instinctive faculty.

We may be the more excused for this preference when we weigh the interest of the genus *Apis* to man. The wasp boots us nothing, but is the pilferer of our fruits, and a marauder upon the hive, whose inhabitants it destroys and consumes their produce, it being indifferent to them which they obtain—the bee or the honey,—either furnishing them with sustenance. The ant is obtrusive and incommodious, making incursions upon the pantry, the store-room, the green-house, and the hot-house; disfiguring our flower-beds, and often disgusting us with our aliment by the impertinent intrusion of its appearance. But the bee stores up for us honey, whose cruses are as inexhaustible as the oil cruse of the good widow of Zarephath, and whose waxen shards furnish us with a beautifully soft light, which in Catholic worship adds solemnity to the rites of religion. In doing this the bee fulfils a sovereign function in the economy of nature, by the fertilization of the flowering plants, with which she reciprocates benefits; the preponderance, however, is importantly in favour of the flower.

If captious objectors should dispute the position we thus claim for the bees, we will willingly leave them the wasp with its sting, whilst we sedulously cultivate the active and industrious bee, whose associations range through all the fields of poetry, but nowhere more lusciously than in the beautiful compositions of the Sanskrit poets Kalidasa and Yayadeva.

The position of the family, whose English constituents I shall subsequently treat of, being thus fixed, I have next to explain the several subdivisions into which it is divided in the following arrangement.

I am prompted to propose this new distribution of the British bees, by the manifest imperfection of the several arrangements of them already extant. The defects of these systems I shall have occasion to exhibit in reference to the course I have been induced to take.

Mr. Kirby's keenness of observation led him to surmise, from the absence of polliniferous brushes upon the posterior legs, or other parts of the body of some, that there might be a class of bees analogous to the cuckoo, amongst the birds, who did not rear their own young, or undertake any of the cares of maternity; but that led by a peculiar instinct they deposited their eggs in the nests of more laborious kinds, for their young to be nurtured upon the provision laid up in store by the latter for the supply of their own progeny. This being merely a supposition, Mr. Kirby made no use of it in the distribution of his families.

Observation has since confirmed the conjecture, and the fact lends material aid to the combination of the bees into detached groups, and which has been partially applied since by all systematizers.

Conjunctively with the assistance derived from this circumstance, the various modes whereby pollen is collected and conveyed, either on the legs or on the belly, further facilitates the grouping of the family. Other structural or economical peculiarities lend their aid, and although the arrangement primarily emanates from the differences in the formation of the tongue, these are corroborated by differences in other organs, and the general distribution, as well as the special combinations, all result from natural characteristics.

The simplicity of the arrangement thus effected is very striking; and we thus find all the bees having similar habits, and with a similar structure united together by it in distinct groups.

I will here insert my scheme, and exhibit why and in what it differs from those of my predecessors; and, where necessary, I shall append such observations upon the several methods extant, as will sufficiently show the necessity, and vindicate the introduction of a new one.

Family MELLICOLLIGERÆ (Honey-collectors).

Subfamily 1. ANDRENIDÆ (Subnormal Bees).

Section 1. *With lacerate paraglossæ.*

Subsection *a*. WITH EMARGINATE TONGUES.

Genus 1. COLLETES.
„ 2. PROSOPIS.

Subsection *b*. WITH LANCEOLATE TONGUES.

Genus 3. SPHECODES.
„ 4. ANDRENA.
„ 5. CILISSA.

Section 2. *With entire paraglossæ.*

Subsection *c*. WITH ACUTE TONGUES.

Genus 6. HALICTUS.
„ 7. MACROPIS.
„ 8. DASYPODA.

Subfamily 2. APIDÆ (Normal Bees).
Section 1. *Solitary.*
Subsection 1. SCOPULIPEDES (brush-legged).
a. *Femoriferæ* (collectors on the entire leg).
† *With two submarginal cells.*

Genus 9. PANURGUS.

b. *Cruriferæ* (collectors on the shank only).
† *With two submarginal cells.*

Genus 10. EUCERA.

†† *With three submarginal cells.*

Genus 11. ANTHOPHORA.
„ 12. SAROPODA.
„ 13. CERATINA.

Subsection 2. NUDIPEDES (naked-legged).
a. *With three submarginal cells.*

Genus 14. NOMADA.
„ 15. MELECTA.
„ 16. EPEOLUS.

b. *With two submarginal cells.*

Genus 17. STELIS.
„ 18. CŒLIOXYS.

Subsection 3. DASYGASTERS (hairy-bellied).
All with two submarginal cells.

Genus 19. MEGACHILE.
„ 20. ANTHIDIUM.
„ 21. CHELOSTOMA.
„ 22. HERIADES.
„ 23. ANTHOCOPA
„ 24. OSMIA.

Section 2. *Cenobites* (Dwellers in Community).
  Subsection 1. SPURRED.
  † *Parasitical.*
      Genus 25. APATHUS.
  †† *Collectors.*
    Temporarily social.
      Genus 26. BOMBUS.
  Subsection 2. UNSPURRED.
    Permanently social.
      Genus 27. APIS.

The primary division of the bees into two large branches, viz. into the *Andrenidæ*, or abnormal bees, and the *Apidæ*, or normal bees, is effected by the mode in which they fold the cibarial apparatus in repose. In the description of the structure of the imago, I have enlarged upon these organs, and for their explanation I must refer to that chapter where diagrams exhibit the structure of the different kinds of trophi of the bees, as well as their mode of folding. Here it is only necessary to notice that in the *Andrenidæ*, the joint at the base draws back the basal portion when protruded, and this basal portion is further jointed at the point of the insertion of the paraglossæ and labial palpi, and parallel with which joint the maxillæ are likewise jointed close to the sinus where the maxillary palpi are inserted laterally upon it. The basal portion thus throws the anterior part forward or retracts it, at the will of the insect, and in the latter case, being then in repose, it lies in contiguous parallelism to the basal half, but beneath it. When thus withdrawn, the short tongue itself, with its paraglossæ and labial palpi are sheltered beneath the coping of the labrum and the lateral protection of the mandibles, whilst the horny sheathing of the maxillæ protect the softer parts folding underneath.

In the *Apidæ*, or normal bees, the basal joint has the same action in withdrawing the entire organ into its place of rest; but the joint which gives it this power is not in an analogous situation to that in the *Andrenidæ*, for it is seated short of the joint which lies at the base of the several organs of the cibarial apparatus. By bending these downwards, it carries their apex backwards towards the basal fulcrum through the action of these two joints, and, when there, the more delicate ones are protected from abrasion or injury, by the lateral overlapping of the horny skin of the maxillæ. All being thus withdrawn within this covering, upon the joint which folds them back, seated at the base of the tongue, the labrum falls, and further to strengthen this protection, the mandibles close over it like forceps.

That this difference in the arrangement of the cibarial apparatus points to any distinctive peculiarities of economy has not been ascertained, for the habits of the *Scopulipedes* greatly resemble those of the *Andrenidæ*; although the habits of one of them, *Anthophora furcata*, are remarkably like those of the foreign genus *Xylocopa*, in its mode of drilling wood. But the *Apidæ* have cross affinities amongst themselves, thus *Ceratina* resembles *Heriades*, and some of the *Osmiæ*, in the way in which it nidificates.

The tongues of the *Andrenidæ* are always shorter, broader, and flatter than those of the *Apidæ*, in which they are always long, cylindrical, and tapering. In the first section of the *Andrenidæ*, the paraglossæ are obtusely terminated at the apex, thence called lacerated, and where they are fringed with brief bristles. The peculiar form of the tongue in this section suggests its being separated into two subsections, that organ being in the first subsection very broad and bilobated, which gives those insects their position in the series by approximating them to the preceding family of the *Diploptera,* or wasps, whose tongues have the same bilobate form, but each lobe in them is furnished with a gland. These tongues, in both cases of the wasps and these bees, may conduce to the building or plastering habits of the insects. The form may aid the wasp and the *Colletes,* the first in the moulding of its hexagonal *papier-mâché* cells, as it may the second in shaping and embroidering the silk-lined abode of its embryonic progeny. Why *Prosopis* should have this organization is difficult to conceive, unless it be from an analogy of structure incidentally previously referred to, beyond which any special object has hitherto escaped detection.

In the second section of the *Andrenidæ,* which have the paraglossæ entire and terminating in a point, the tongues all also terminate acutely with a lateral inclination inwards. In the lanceolate-tongued tribe they bulge outwards laterally, although pointed at the apex.

All this subfamily of *Andrenidæ,* excepting only the two genera reputed parasites, viz. *Prosopis* and *Sphecodes,* are essentially *Scopulipedes,* densely brush-legged, for the conveyance of pollen which they vigorously collect; but from the brevity of their tongues they are restricted to flowers with shallow petals and apparent nectaria, their favourite plants being the abounding *Compositæ* and *Umbelliferæ,* as well as the *Rosaceæ,* whence they derive the agreeable odours which many of them emit upon being captured.

Their peculiar mode of collecting is a further reason for bringing the brush-legged *Apidæ* collectively to the top of the normal bees, in juxtaposition to the *Andrenidæ*, where the transition is made very naturally from *Dasypoda* to *Panurgus*.

The whole of the cibarial apparatus, or trophi, is always complete in all its constituent parts throughout the *Andrenidæ*; and it is only with *Ceratina*, in the group of *scopuliped Apidæ*, that it begins to show the tendency it has to abnormal deficiencies, by the paraglossæ, in that genus, being obsolete. This characteristic, then, exhibits itself in the *Nudipedes* with two submarginal cells who are parasitical upon the *Dasygasters*, in whom also the maxillary palpi participate in a deficiency in the authentic number of their joints, whilst in *Apis* both maxillary palpi and paraglossæ are unapparent. This shows that the numerical completion of the organs of the mouth have nothing to do with the qualifications of the creature, the best endowed in other respects being thus curtailed, the final cause of which is not yet understood.

The shape of the tongue itself thus separates the *Andrenidæ* into three well-defined divisions readily perceptible. These, as I have just observed with respect to the differences in the mode of closing the oral apparatus in both cases, yield no clue to economy and habits, for which observation must supervene to illustrate it. This, patiently carried out, is very desirable, as it is still in discussion whether, notwithstanding the elucidation structure affords, *Prosopis* and *Sphecodes* are or are not parasitical. Structure says they are, for, like the cuckoo-bees forming the group *Nudipedes* in the *Apidæ*, they are destitute of the requisite apparatus for collecting pollen. Mr. Kirby, however, gives direct testimony in favour of *Sphecodes* being a burrower, in the case of which bee it ought not to be a matter of much difficulty to determine, for on sandy plateaus I have occasionally found it very abundant, especially where there was ragwort (*Senecio*) in flower in the vicinity, to which the males resorted; but being at the time more intent on other matters, I neglected the opportunity. Other observers concur with Mr. Kirby as regards *Sphecodes*, and also say as much for *Prosopis* (better known as *Hylæus*). I strongly incline to the opinion enunciated by Latreille and Le Pelletier de St. Fargeau, that they are parasites. My opinion is based upon peculiarities in them other than, although strengthened by, the negative characteristic of absence of polliniferous organs. A negative cannot be proved, it is true, yet what has been positively asserted may as certainly result either from defective observation, or from too strong a desire to find no parasites among the *Andrenidæ*. My reasons occur elsewhere in this work, and I need not repeat them. It is still an open question, and the young entomologist, if entering the arena unprepossessed, might win his spurs in determining it. It would be well worth the trouble of attending to for those who have leisure, and if decided in favour of the independency of these genera, which must be corroborated by a plurality of observations, and not confined to one locality, they would form strong and remarkable instances of a defective analogy in nature's workmanship, and suggest looking further for the causes of so extraordinary an anomaly, and urge us to endeavour to trace the equivalent which supersedes it.

The main subdivision of the *Apidæ* results from the habits of the insects, which divides them into SOCIAL and SOLITARY. The only tangible characters the social tribes present to distinguish them from the solitary is the glabrous surface of the posterior tibiæ, with their lateral edges fringed with bristles slightly curved inwards, and which form, with the slightly indented surface of the limb, a sort of natural basket for the conveyance of pollen or other stores to the nest. This, however, has not been made use of as a main feature for scientific distribution, although they might follow the *Dasygasters*, as *corbiculated bees*, or little basket bearers, in which case they would form as pertinent a group as any of the rest, and the whole distribution of the bees, *Apidæ*, would then rest upon the absence of, or the mode in which the polliniferous organs were present. But the wonderful attribute of their extraordinary instinct prohibits their being treated with the rest in a consecutive line, and renders it rationally imperative that all the *Cenobites* should group together in a section by themselves, and separate from the rest. Therefore in my arrangement I have not availed myself of this very natural character, and here indicate it, to show that I have not passed it from not noticing it.

Although the division into social and solitary yields in itself no tangible character whereby the insects may be separated, it being wholly empirical, yet is it so natural and necessary that it is impossible to gainsay it. We find the solitary section readily resolve itself into groups or subsections, determined by positive structural characters, indicative of certain habits, and having a conforming economy, besides which they are equivalents.

Thus the first subsection presents us with the brush-legged *Apidæ* (*Scopulipedes*), which collect pollen upon their posterior legs. These are further subdivided into those which collect it upon the whole limb, viz. the coxa, the femur, the tibia, and first joint of the tarsus, (the *femoriferæ*), and those which gather it merely upon the shank and basal joint of the foot (the *cruriferæ*). These collectively form a well-defined group, and why *Panurgus* should be separated from the brush-legged bees, when it is a most conspicuous instance of the faculty, even more so than any other of the *Scopulipedes*, I have yet to learn. It is true its mode of collecting closely resembles that practised by the *Andrenidæ*, as does also the furniture for the purpose of its posterior legs, but being essentially collocated with the *Apidæ* or normal bees by its tongue, it fittingly links itself to the other brush-legged *Apidæ* (which have hitherto been placed between the *Dasygasters* and the Social Bees), by means of the genus *Eucera*, by reason of its two submarginal cells, the structure of its maxillary palpi, its mode of burrowing, and by each being infested by a similar parasite— a *Nomada*, which in accommodation to the size of the sitos is the largest of the genus. *Nomada* does not occur as a parasite upon any other of the brush-legged bees, or indeed upon any other of the true bees at all, which peculiarity brings these two genera into close contiguity to all non-parasitical *Andrenidæ*, all of which have their legs furnished with polliniferous brushes, and upon which subfamily, exclusively of these two instances of *Panurgus* and *Eucera*, *Nomada* is solely parasitical.

With respect to the two submarginal cells to the wings, nature must have some reason for the limitation, for we find it prevalent also throughout the *Dasygasters*, or hairy-bellied bees.

The next very natural group is consistently central. It comprises the cuckoo-bees, which are naked-legged

(*Nudipedes*), by reason of their parasitism, they not requiring organs to collect what they have no occasion to use. Their parasitism extends both upwards and downwards, those with three submarginal cells being parasitical upon all the brush-legged bees, whether subnormal *Andrenidæ* or the *Scopulipedes*, those with two submarginal cells being restricted in their parasitism to the *Dasygasters*.

These *Dasygasters*, or hairy-bellied bees, form the next very natural group. Their general peculiarity of structure I have had occasion to advert to, in treating, in a former section of the work, upon the structure of the imago, and to which I now refer to avoid repetition. This group contains the majority of the artisan bees, whose habits I shall particularize when I speak of the genera specially; but we find carpenters amongst the *Scopulipedes*, and essentially builders amongst the *Cenobites*, which form a further and the last of our natural groups. A true cuckoo-bee (*Apathus*) consorts amongst these *Cenobites*, and properly so, from many causes. The anomaly would have been too great to have removed it to a place amongst the *Nudipedes*, for although in obsolete paraglossæ, and in a deficiency in the normal number of the joints of the maxillary palpi, it resembles some of these, its general habit and general structure, bating that controlled by its parasitical habits, are so like *Bombus*, that it cannot well be separated far from the latter,—especially as we know too little of its habits to say that it does not regularly dwell in the nest of its sitos, which may well mistake it for one of its own community, it resembling the species it infests so closely; it therefore consistently associates systematically with the temporarily social societies.

Having thus cursorily skimmed the surface of the method I suggest, I have next to give my reasons for proposing it in lieu of adopting any yet extant.

My exhibition of Kirby's grouping, in the preceding section, where I treat of the scientific cultivation of British bees, will fully explain why I could not adopt that arrangement.

Why I cannot follow Latreille's, is, that in his last elaboration, in his 'Familles Naturelles,' published in 1825, which must be considered as his final view, he does not satisfactorily divide the *Andrenidæ*, of the genera of which he has made a complete jumble. With the *Apidæ* in his group of *Dasygasters*, he intermixes *Ceratina*, separating it from the group of *Scopulipedes*, where it truly belongs by every characteristic, and he mingles also with them the two cuckoo genera *Stelis* and *Cœlioxys*, which are merely parasites upon these *Dasygasters*, and can only be associated by the structural conformity of the two submarginal cells to the superior wings, and the length of the labrum, the latter being a character of very secondary importance; and further, he dissevers the *Scopulipedes* in placing *Panurgus* at the commencement of the *Apidæ*, and the rest proximate to the social bees.

Westwood, in his modification of Latreille's system, certainly divides the *Andrenidæ* better than his master had done, but he does not go far enough. Besides, he interposes *Halictus* and *Lasioglossum*, (the latter admitted as a genus merely out of courtesy to Curtis, who had elevated it to that rank in his 'British Entomology,' although it is nothing more than a male *Halictus*), between *Sphecodes* and *Andrena* with *Cilissa*, these having lanceolate tongues with lacerate paraglossæ, whereas *Halictus* has a very acute tongue, and its paraglossæ are entire, as is also the case with *Dasypoda*, from which *Halictus* is thus divided. In the *Apidæ*, he does not separate the cuckoo-bees, but with Latreille intermixes *Cælioxys* and *Stelis* with the artisan-bees, although without retaining Latreille's convenient and suitable name of *Dasygasters*, for this group of mechanics. The same objection I take to his *Scopulipedes* as that expressed above, relative to Latreille's.

Precisely the same fault I find with the *Andrenidæ* of Smith, as that urged above with respect to Westwood's. He is more careful with his *Apidæ*, his *Cuculinæ* being all genuine parasites, but he includes *Ceratina* with the *Dasygasters*, with which it has no affinity of structure, and only a slight analogy in the *form* merely of its abdomen without its hairiness beneath, to that of *Osmia*, from whose proximity he takes it to place it near *Heriades*, when it is certainly intimately allied in every respect with the *Scopulipedes*, and by reason of its subclavate antennæ might suitably be brought into juxtaposition with *Panurgus*, did not its obsolete paraglossæ and three submarginal cells interfere with its occupying this position. To his *Scopulipedes* the same objection is valid as that taken to Latreille's and Westwood's disposition of them. Amongst the social bees he separates *Bombus* from *Apis*, by the intervention of *Apathus*, which is scarcely consistent.

It is in no spirit of captiousness that these objections are made; they are deduced from collocations whose conspicuous incoherence is patent to the most superficial observation. The distribution I have here introduced has been made merely to ameliorate, and make more cogent, what was so palpably defective and feeble.

CHAPTER IX.

A TABLE, EXHIBITING A METHOD OF DETERMINING THE GENERA OF BRITISH BEES WITH FACILITY.

The following table is constructed exclusively to facilitate, by the most obvious characters, the recognition of the several genera into which the family is divided; it will, however, be incumbent upon the learner to use some diligence in order to acquire an accurate perception of their distinguishing characteristics.

By the present extremely artificial plan the systematic sequence is disturbed; but the numbers, which will be found appended to the names in the table, will show their orderly succession.

The natural generic character which precedes the account of each genus in the next division of the work will give the reason, by comparison, of the order in which "system" arranges them, and which being based mainly upon the differences of the trophi,—although, conjunctively with other characters, the trophi must necessarily be studied for its explanation,—their description in the description of the part of the imago is consequently referred to.

Did we know exactly the uses of the component parts

of the trophi severally, we should be better able to determine the legitimacy of applying them to the purpose of indicating the natural generic character, but being compelled, by reason of our ignorance of their several special functions, to avail ourselves of their form, relative proportions, and number only, uncertainty of having caught the clue of nature's scheme must of necessity attend this distribution.

But as what we do know of their uses in this family clearly indicates them to be an essential instrument indispensable to the economy of the insect, and which gives these organs an almost paramount importance, their comparative construction in the several genera would yield clear notions of the true order of succession, were we acquainted with the relative significancy of the various portions of the entire organ. Thus we see it numerically most complete in what we are pleased to suppose the least genuine bees—the *Andrenidæ*.

In my series of the genera proposed in the preceding section, with the *Nudiped* true bee *Melecta* commences a deficiency of either some of the joints of the maxillary palpi, or of the paraglossæ;—throughout the artisan bees this abridgment is conspicuous both in number and proportion; and it culminates in what we consider the *facile princeps*, that most wonderfully organized of all insects—the genus *Apis*, which in its neuters has neither paraglossæ nor maxillary palpi, the latter being equally deficient in the male or drone, and in the queen; and in both the male and the queen the paraglossæ are but rudimentary.

Nature appears too mysterious in her operations to permit us to solve these remarkable anomalies, for no combination of the genera founded exclusively upon them supplies us with Ariadne's thread. Every such combination breaks up more harmonious groups, and we then retrace our steps, satisfied that we are on the wrong road.

In some other orders of insects the cibarial apparatus has but little bearing upon the insect's mode of life, for in many it is not used either for nutrition or in their economy, or so slightly so as to admit of its being considered of very inferior importance, although systematists—to enhance the value of their own labours, by the frequent difficulty, from excessive minuteness, of its examination—have usually made it a prominent feature in their arrangements.

That science has not widely strayed away from the true succession and natural affinities by the main selection of the trophi for the arrangement of the bees, seems partially confirmed by the gradations of form or habit that this method of treatment in general exhibits. A higher method doubtless exists, which would give form, number, and proportion very inferior rank in ordering the arrangement, but at present the clue to it has not been discovered.

These questions are indeed beyond the scope of a work of this character, which is merely a ladder to the fruits of learning, and the bearing of them is only hinted at to indicate that there is much exercise for the intelligence in the study of even this small family. The mind that would stop in the study of nature at the knowledge of genera and species, can be very speedily satisfied, and one bright spring day's successful collecting will furnish the materials for much patient and industrious occupation.

In nature we find all things apparently blended in the

grandest confusion; but they all have mutual and reciprocal bearings which give a definite purpose to the seeming disorder, and which make each separate unit the centre of all. But we, from our inability to grasp in its fulness the order of this disorder, are obliged to seize fragments and, separating them into what we conceive to be their coherent elements, use them as exponents of the entirety. They could not so exist in nature, but would speedily die out, and it is only by the way in which we find them intermingled, that they can be maintained. Thus, as all conduce to the conservation of each, each conduces to the conservation of all.

A large collection of natural history, composed of every available item that can be gathered from every kingdom of nature's vast domain, may perhaps be compared (*magnis componere parva*) with the constituent parts of a most elaborately-constructed and complicated clock, which its skilful artificer has designed and made to record and chime the divisions of time, and to register the days, weeks, months, and seasons, and which a virtuoso having taken to pieces, has sorted into its details of wheels and springs, levers and balances, chains, bells, and hands, which told the time when its music would peal; and arranging like to like, thinks he will thus understand more clearly the complexity of the varied movements. But, sadly disappointed, he finds he cannot comprehend the combination of the intricate machinery, although he singly admires the minute perfection of each delicate and ingenious piece lying before him which composed the structure, but which has now lost all expression, his curiosity having deprived the organism of its vitality, which is its most wonderful element.

And this is our process, for if we stop here we have

but an assortment of vapid machinery, no click of whose wheels gives note of the vital hilarity of their relative and combined effects. The final cause of creation escapes us thus frittering it into details, which if we merely abide by, we but loiter at the foot of Pisgah, instead of ascending its summits to survey thence the sunny and varied landscape, the glorious sea, and, arching over all, the blue cope of heaven. The manifold relations of animate and inanimate nature, which, although they must be studied in detail, are to be appreciated in their entirety, should stimulate the efforts of the naturalist to conquer all impending difficulties, and he should not permit himself to be satisfied with this preliminary knowledge.

Although the above be the inevitable effect of distributing nature into its component parts, it is the indispensable precursor to the study, for the scientific treatment is the only mode whereby, through special study, we can arrive at the comprehension of the great generality. We thus strive to trace the mode in which each emanates from each; and even when this is not absolutely tangible we may discover affinities or analogies by structural resemblances which implicitly lead to physiological inferences, and thence on, higher and higher, all lending us aid to make the larger survey, wherein we behold the concatenation of the many links which harmonize the spiritual with the material. But the study must be thorough, and its details are not to be spread out before us merely as a beautiful picture-book. They all have their place in the great ordinance of nature, which it is for us to find. At first we can only spell the syllables, which the study of species puts together for us, but by degrees we shall trace the words, and read the sentences : a study more abstruse but far more pregnant than that of the Egyptian hieroglyphics, and whose attainment is rewarded with a supremer knowledge than is accorded by these, which exhibit merely the legends of dead despots; but here we have a display of the vitality of the wisdom inscribed in gleaming characters upon the leaves of the wonderful book of life, God's glorious works, made manifest to man.

Thus we should aim at the knowledge of final causes, the apparent wisdom of whose adaptations points clearly to the source of all—the first great Cause. A naturalist with such large views has a wide field before him, which with every step expands, and which alone is worthy of engrossing the earnest attention of his intelligence, and is in itself sufficient to absorb the profoundest contemplation. His mind becomes thus filled with great objects, which charm it with their beauty and feed it with the complexity of their intricate combinations, whose earnest development is an affluent stream of perpetual instructive occupation. With Newton we may say: "We everywhere behold simplicity in the means, but an inexhaustible variety in the effects," resulting all from the luminous wisdom of prearranged design.

The humiliation which attends the sentiment of the utter inability and incompetency of the mind to grasp the intricacy and vastness of nature, is consoled by the redundant proofs the contemplation yields of a supreme and benevolent Providence presiding over all things, and thence we derive the comfortable and supporting assurance, in the fickle waywardness and vicissitudes of a harassed and anxious life, that a benevolent eye is ever watchfully awake; for the naturalist everywhere beholds that omnipotently wise and loving Providence in active operation throughout nature.

No study like natural history, pursued in a humble and docile spirit, so harmoniously elicits the religion of the soul, or than which so fitly prepares it to enter, by the pathway of the works of God, the august temple of His revealed Word.

But to return: what we call science is the mere accidence of nature, which in fact aggravates our infirmity by permitting our intelligence to attempt to grasp, through the various details, their intricate combinations. But as truth sooner arises out of error if methodically pursued, and its results recorded, than out of confusion and guesswork, theories based upon observation, however inaccurate at first, ultimately lead up to the certain acquisition of the truth itself.

AN EASY DISTRIBUTION OF THE BEES,

THE NUMBERS REFERRING TO THE SCIENTIFIC SERIES.

ANDRENIDÆ (Subnormal Bees).

TONGUE SHORTER THAN THE MAXILLÆ, PORRECT.

Posterior tibiæ clothed with hair to convey pollen.

Two submarginal cells.
    Posterior legs very robust, polliniferous hair on tibiæ and plantæ dense but short . . . . . . MACROPIS (7).
    Posterior legs slender; polliniferous hair on femora, tibiæ, and plantæ dense and very long . . . . . DASYPODA (8).

Three submarginal cells to the wings.
   Abdomen truncated at base . . . Colletes (1).
   Abdomen ovate.
      Abdomen entire at apex; maxillary palpi as long or longer than the maxillæ . . . . Andrena (4).
      Abdomen entire at apex; maxillary palpi half the length of the maxillæ . . . . Cilissa (5).
      Abdomen with a vertical incision at the apex . . . . Halictus (7)

Posterior tibiæ without hair to convey pollen.
Two submarginal cells to the wings . Prosopis (2,.
Three submarginal cells to the wings . Sphecodes (3).

APIDÆ (Normal Bees).

TONGUE AS LONG OR LONGER THAN THE MAXILLÆ, INFLECTED BENEATH, AND COVERED BY THE MAXILLÆ IN REPOSE.

Without polliniferous organs.

Two submarginal cells to the wings.
   Abdomen at apex rounded . . . Stelis (17).
   Abdomen at apex conical . . . . Cœlioxys (18).
Three submarginal cells to the wings.
   Abdomen lanceolate . . . . . Nomada (14),
   Abdomen subtruncate at base.
      Abdomen obovate, thorax glabrous Epeolus (16).
      Abdomen subconical, thorax hirsute . . . . Melecta (15).
      Entire body densely hairy . . Apathus (25).

With polliniferous organs.

Pollen conveyed on the venter.
   Two submarginal cells to the wings of all.

Abdomen subclavate.
    First three joints of labial palpi continuous, terminal joint inserted before apex of third . . CHELOSTOMA (21).
    First two joints of labial palpi continuous, two last inserted before the apex of the second . . HERIADES (22).
Abdomen obovate, rounded at apex   OSMIA (24).
Abdomen truncated at base.
    Segments slightly constricted, and not spotted with colour . MEGACHILE (19).
    Segments not constricted, spotted with yellow . . . ANTHIDIUM (20).

Pollen conveyed on the posterior legs.
    Two submarginal cells to the wings.
        Abdomen lanceolate; antennæ clavate; posterior legs covered with long hair . . . . . . PANURGUS (9).
        Abdomen obovate; antennæ filiform; posterior legs covered densely with short hair . . . EUCERA (10).
    Three submarginal cells to the wings.
        Short dense hair on the whole posterior tibiæ externally.
            Abdomen obovate; first joint of labial palpi twice as long as second . . . . ANTHOPHORA (11).
            Abdomen subrotund; first joint of labial palpi six times as long as the rest . . . . . SAROPODA (12),
        Long hair, but loose, on the entire posterior tibiæ, externally and internally.
            Abdomen subclavate . . . . CERATINA (13).
        Curved hair fringing the edge only

of the posterior tibiæ, the
centre glabrous.
Body densely hirsute, spurs to
all the tibiæ . . . . . . Bombus (26).
Body subpubescent, no spurs
to the posterior tibiæ . . . Apis (27).

It will be desirable to add a few observations to the preceding table to facilitate its use, and because, as many of the characters upon which it is framed are exclusively those of the female, it is necessary to point out the differences of their males, that the sexes of the genera may be duly recognized and associated.

It may be first noticed generally that the antennæ, in the males, are not usually geniculated at the scape, which is nearly always the case in the opposite sex, and they are also, with rare exceptions, always longer than those of their females. In *Colletes, Prosopis, Dasypoda, Panurgus, Ceratina, Nomada, Melecta, Epeolus, Stelis,* and *Anthidium,* the habit or colouring of the males is so similar to that of the females, that their genus may be thus at once determined, and, in fact, the brief characters in the table will embrace them.

The male *Eucera* can be distinguished from those of *Anthophora* and *Saropoda*, both by the differences in the number of the submarginal cells of the wing, and by the extreme length of its antennæ, whence the genus derives its name. In *Andrena* and *Cilissa*, the males have usually lanceolate bodies. In the latter genus there will be no difficulty in associating the legitimate partners; but in *Andrena*, although general habit will usually bring the male within the boundary of the genus, nothing but experience, or specific description will associate the sexes correctly, there being in many cases an extraordinary discrepancy between them. These two genera themselves also can scarcely be distinguished apart, excepting by means of their trophi; *Cilissa*, however, in general habit greatly resembles the genus *Colletes*, especially the *Cilissa tricincta*, which might, upon a superficial glance, be almost mistaken for one of them.

The male *Halicti* have long cylindrical bodies and long antennæ, but from the male *Chelostoma*, which has a very similarly shaped body also and long antennæ, they may be distinguished by the differences in the number of the submarginal cells; and from those of *Sphecodes*, by the antennæ, which, in the latter are not relatively so long, and are usually moniliform. The thorax of these is also less pubescent, and the tinge of the red colour of their abdomen is different from that of the red male *Halicti*.

The males of *Cælioxys* can be readily distinguished from those of *Megachile*, by the spinose apex of their abdomen. In *Megachile*, general habit will bring the males within the precincts of their genus, as well as their largely dilated anterior tarsi in some of the species.

A difficulty similar to what is found in the distinction between *Andrena* and *Cilissa*, arises in the separation of *Chelostoma* from *Heriades*, and which we shall again meet with in drawing the line between *Anthophora* and *Saropoda*. The difference can only be detected by examining the trophi, but a pin and a little patience will elucidate the separation. The males in all but two species of *Anthophora* may be readily associated with their partners; but in these two the females are entirely black, and so hirsute as to have led Ray (wanting the knowledge of the use of the trophi and posterior shanks) to unite the one he knew with his *Bombylii;* their males are fulvous, and the latter have a remarkable elongation of the intermediate tarsi, from one of the joints of which also a tuft of hair or a loose lateral fringe projects, giving them thus a wider expansion, and the use of which is prehensile, the same as that for which the anterior tarsi in some of the *Megachiles.* and in our single *Anthidium* receive their dilatation. This structure has also the effect of adding very considerably to the elegance of their appearance when they are in fine condition.

The male *Apathi* can only be distinguished from the male *Bombi* by familiarity with specific characteristics, or by the examination of the trophi. But the former is the more certain mode of separation, as the trophi in *Bombus* vary in some species, but not sufficiently to authorize generic subdivison. General appearance will mark where they approximately belong. The length of their antennæ sufficiently distinguishes them as males, and they may be taken with impunity in the fingers from flowers for examination, being, like all the male aculeate *Hymenoptera,* unarmed with stings. The female *Apathi* may be superficially distinguished from the female *Bombi,* which they most resemble, exclusively of the generic characters of the convex and subpubescent external surface of the posterior tibiæ and the trophi, also by their abdomen being considerably less hirsute than that of the genuine *Bombi,* in which it is entirely covered with dense shaggy hair, whereas in *Apathus* there is a broad disk upon its surface nearly glabrous. If I remember rightly, it is the male *Apathi* only, and not the male *Bombi,* which emit on capture a pleasantly fragrant odour of attar of roses.

The table will suffice for distinguishing the male *Apis* from all other male *Apidæ*, and which has a further peculiarity exhibited by no other of our native bees, in the conjunction upon the vertex of the compound eyes, in front of which, upon the frons, the simple eyes or ocelli are placed in a very slightly curved line.

These indications are enough to enable the beginner to work his way smoothly, and a little practice will soon render these observations superfluous.

The economy of nature is so perfect that wherever we can trace a difference, we may assume that a reason and a purpose exist for the variation. Thus we do not know why some bees have three submarginal cells to their wings, and others only two. Nor do we know what governs their variety of shape. The deficiency we might think implied inferiority; but this cannot be, for those with most frequently the smaller number, viz. the artisan bees, are, in the majority of cases, the most highly endowed, and have the most special habits.

In the relative numbers of the maxillary and labial palpi, there are remarkable differences, the reason for which we cannot trace, for, as before observed, we do not know even their function, which would perhaps guide us to other views. Their normal numbers are six maxillary, and four labial palpi. The latter take remarkable relative development and peculiarity of insertion and form, especially in the *Apidæ*; but throughout the whole series of our bees, they are never reduced to fewer than their normal number, whereas the maxillary palpi never have similarly large development of structure, and are variously modified in number and consistency from the typical or normal condition.

Thus in *Eucera* and *Melecta* there are but five joints;

in *Osmia* and *Saropoda*, four; in *Chelostoma* and *Cœlioxys*, three; in *Anthidium* and *Megachile*, etc., two; and in *Epeolus* and *Apis* but one.

In this collocation no incidental peculiarity beyond diversity is apparent, for in the first instance a parasite and a bee not parasitical are associated; and in the last, a parasite is associated with the bee which has the most elaborate economy, and the most largely developed instinct of all known insects. Nor are, in any case, those parasites associated by these means with their own sitos, or insect upon which they are parasitical.

Thus encouragement attends the beginner at the very outset of his study; and the prospect of a wide field for discoveries, in many directions, lies open to him, to excite his curiosity and to stimulate his industry to the pursuit of higher aims than the mere accumulation of species.

CHAPTER X.

THE SCIENTIFIC ARRANGEMENT AND DESCRIPTION OF THE GENERA, WITH LISTS OF OUR NATIVE SPECIES AND AN ACCOUNT OF THE HABITS AND ECONOMY OF THE INSECTS, WITH INCIDENTAL OBSERVATIONS SUGGESTED BY THE SUBJECT.

I now proceed to the treatment and description of the genera severally, and the enumeration of the species in due scientific consecutive order.

The generic names adopted are those of the first describers of the genera; but the generic characters given by them could not be employed, they having been usually framed to suit special purposes.

All the generic characters introduced into this work are therefore quite original, and have been made from a very careful autoptical examination of the insects themselves.

The synonymy added to the lists of species is limited to the species described in Mr. Kirby's work, where he is not the first describer, or to those of such other English works wherein the species may have been described in ignorance of its previous registration.

The observations appended, wherein the habits of the insects are described, will be found to embrace discursive subjects suggested by the matter in hand, and here a dry didactic style has been purposely avoided, as in the majority of cases they record the personal experiences or notions of and hints from an old practical entomologist.

Class INSECTA METABOLIA, *Leach.*

Order **HYMENOPTERA**, *Linnæus.*

Division *ACULEATA*, Leach.

Antennæ in male with 13 joints, in female with 12. Abdomen in male with 7 segments, in female with 6.

Family MELLICOLLIGERÆ (Honey-collectors), *Shuck.*

Subfamily 1. Andrenides (Subnormal Bees), *Leach.*

Syn. Genus Melitta, *Kirby.*

The maxillary palpi always six-jointed.

Section 1. *With lacerate paraglossæ.*

Subsection *a.* Linguæ emarginatæ (with emarginate tongues).

Syn. Obtusilingues, *Westw.*

Three submarginal cells to the wings.

Genus 1. COLLETES, *Latreille.*

(Plate I. fig. 1 ♂ ♀.)

Melitta * *a,* Kirby.

Gen. Char.: Head transverse, flattish; *ocelli* in an open triangle on the vertex; *antennæ* not geniculated, but slightly curved, filiform, short; joints, excepting the basal or scape, which is as long as five of the rest and slightly curved, nearly equal; *face* beneath and within the insertion of the antennæ, slightly protuberant, laterally flat or concave; *clypeus* convex, margined anteriorly, entire; *labrum* transverse, slightly produced in the centre in front, and the process rounded; *mandibles* obtuse, subbidentate; *cibarial apparatus* short; *tongue* deeply emarginate and bilobate, the lobes fringed with short setæ; *paraglossæ* half the length of the tongue, abruptly terminating and lacerate, and setose at the apex; *labial palpi* much shorter than the paraglossæ, four-jointed, the joints equal and each subclavate; *labium* about the same length as the tongue, its inosculation acutely angulated; *maxillæ* broad, lanceolate, the length of the tongue; *maxillary palpi* six-jointed, not so long as the maxillæ, the two basal joints the longest, the rest equal, short, and subclavate, the apical one rounded. THORAX subquadrate, very pubescent, the prothorax inconspicuous; *scutellum* transversely triangular or semilunate, *postscutellum* lunulate; *metathorax* abruptly truncated, and densely pubescent, especially laterally, for the conveyance of pollen; *wings* with three submarginal cells and a fourth slightly commenced, the second and third each receiving about their centre a recurrent nervure; *legs* all pubescent, the anterior and intermediate on their external surface chiefly, their *plantæ* also setose; the posterior *coxæ*, *trochanters*, *femora*, and *tibiæ* very hirsute, especially beneath, their *tarsi* entirely setose; *claws* bifid. ABDOMEN truncated at the base, subconical with a downward bias, the *segments* with bands of closely decumbent nap, and the surface of all more or less deeply or delicately punctured; the basal segment in the centre, beneath, with a longitudinal tuft of long hair.

The MALE differs in having the *mandibles* more distinctly bidentate, and in being less densely pubescent, especially upon the legs. In general aspect it is very like its female.

Note. The genus *Cilissa* has, superficially observed,

much of the habit of *Colletes*, particularly in the male of *Cilissa tricincta*.

NATIVE SPECIES.

1. *succincta*, Linnæus, ♂ ♀. $3\frac{1}{2}$–$5\frac{1}{2}$ lines.
   succincta, Kirby.
   fodiens, Curtis.
2. *fodiens*, Kirby, ♂ ♀. $3\frac{1}{2}$–$4\frac{1}{2}$ lines.
   pallicincta, Kirby, ♀.
3. *marginata*, Linn., ♂ ♀. 3–4 lines.
4. *Daviesiana*, Kirby, ♂ ♀. $3\frac{1}{2}$–$4\frac{1}{2}$ lines. (Plate I. fig. 1 ♂ ♀.)

GENERAL OBSERVATIONS.

This genus is named from κολλητης, *one that plasters*, in allusion to the habits of the insects, which will be described below. The female insects themselves have, at the first glance, very much the appearance of the working honey-bee, but they are considerably smaller, and, upon a very slight inspection, they are found to be exceedingly distinct. The respective males of the species are conspicuously smaller than their females, but their specific characteristics are very much alike, and there is some difficulty in separating and determining the species. One strong peculiarity, marking all of them, is that the segments of the abdomen are banded with decumbent, hoary or whitish down, in both sexes, and the determination of the species lies chiefly in the variations of these bands, and in the almost entire absence or conspicuous presence of minute punctures covering the segments. The females are very active collectors of pollen, and return from their excursions to obtain it, very heavily laden to their nests. I am not sure that all the species are not gregarious, to use this term in an acceptation somewhat different from its usual application, for here, and whenever used in entomology, it is meant to signify that they burrow collectively in large communities, forming what is called their metropolis, although each bores its independent and separate tube, wherein to deposit its store of eggs. The males, neither in these insects nor throughout the whole family of the bees, participate at all in the labours required for the preservation and nurture of the progeny, a duty that wholly devolves upon the maternal solicitude of the female,—these males having fulfilled their mission, which is not perhaps restricted to their sexual instinct, but may also be conducive to the grand operation of the family in the economy of nature, viz. the fertilization of the flowering plants, flit from blossom to blossom, and thus convey about the impregnating dust. They may also be often seen basking in the sunshine upon the leaves of shrubs, and thence they become lost or dispersed or the prey of their many enemies,—birds or insects, which are always on the alert in search of ravin.

The aspect selected by the females for their burrows, varies according to the species. Some choose a northern, and others a southern aspect; thus, the *C. succincta* seems to prefer the former, and the *C. fodiens* the latter, as does also the *C. Daviesana;* and where they burrow they congregate in enormous multitudes. The mortar interstices of an old wall, or a vertical sand-rock, which, from exposure, is sufficiently softened for their purpose, are equally agreeable to them; nor have they any objection to clay banks.

In these localities each individual perforates a cylindrical cavity, slightly larger than itself, and which it excavates to a depth of from eight to ten inches, or even sometimes less. Now comes into operation the use of the peculiarly-formed tongue with which nature has furnished them, and described above in the generic character. These cells are occupied by a succession of six, or eight, or even sometimes no more than two, three, or four cartridge- or thimble-like cases, in each of which is deposited a single egg with a sufficiency, taught the creature by its instinct, of a mingled paste of honey and pollen, for the full nurture and development of the vermicle that will proceed from the egg upon its being hatched, and wherein this larva, having consumed its provender, becomes transformed into the pupa, and by the continuance of nature's mysterious operations, it speedily changes into the perfect insect. But the beauty with which these little cells are formed transcends conception. Each consists of a succession of layers of a membrane more delicate than the thinnest goldbeater's skin, and more lustrous than the most beautiful satin. In glitter it most resembles the trail left by the snail, and is evidently, from all experiments made, a secretion of the insect elaborated from some special food it consumes, and by means of its bilobated tongue, which it uses as a trowel, it plasters with it the sides and the bottom of the tube it has excavated to the extent necessary for one division. As this secretion dries rapidly to a membrane it is succeeded by others, to the number of three or four, which may be separated from each other by careful manipulation. It then stores this cell, deposits the egg, and proceeds to close it with a coverele of double the number of membranes with which the sides are furnished, and continues with another in a similar manner, until it has completed sufficient to fill the tubular cavity, which, after closing the last case similarly to the rest, it stops up the orifice with grains of sand or earth. The food stored up is subject to fermentation, but this does not appear to be prejudicial to the larva, which first consumes the liquid portion of the store and then drills into the centre of the more solid part, and continues enlarging this little cylinder until increasing in growth by its consumption, it itself fills the cavity, and thus supplies the lateral stay or prop which, by means of the stored provender, was previously prevented from falling in. It has not been ascertained what number of eggs each insect lays, or whether it bores more than one tube, but it is presumable that it may do so, and possibly thus, from the numbers annually produced, for there are two broods in the year, colonies are thrown off which gradually form another metropolis somewhere in the vicinity, although the majority continue to occupy the old habitat from year to year. But the number of these insects is kept within due limits by the individual abundance of the parasites that infest them, and by the unsparing and unflinching attacks of earwigs, which consume all before them,—perfect insect, larva, and provender. The two most conspicuous parasites they have, are the beautiful little bee, *Epeolus variegatus*, the young of which is sustained, as in all bee parasitism, by consuming the food stored for the sustenance of the young of the *Colletes;* and the other is the little dipterous *Miltogramma punctata*, whose larva, evolved from the egg deposited in the cell, feeds upon the larva of the *Colletes*, or possibly upon that of the *Epeolus*, which otherwise would seem to have no check to its fertility, excepting that it may be subdued by the *Forficulæ*.

These insects are to be found during the spring and summer months, and throughout the southern counties, although some species are extremely local. Some occur also in the north of England and in Ireland. I am not prepared to say what flowers they prefer, for I have never captured them on flowers, but they have been found frequenting the Ragwort, and Curtis took a species at Parley Heath, in Hampshire, on the Bluebell (*Campanula glomerata*). They form a remarkable instance of an artisan bee, but so only in its habits, amongst the *Andrenides*.

Two submarginal cells to the wings.

Genus 2. PROSOPIS, *Fabricius*.

(Plate I. fig. 2 ♂ ♀.)

MELITTA * *b*, Kirby.—HYLÆUS, Latreille.

Gen. Char.: HEAD transverse, flattish; *ocelli* in an open triangle on the vertex; *antennæ* geniculated, the basal joint of the flagellum as long as the second, and both subclavate, the rest of the joints short and equal; *face* flat, slightly protuberant between the insertion of the antennæ, and distinguished from the clypeus by a suture; *clypeus* transversely quadrate, slightly widening gradually to the apex, marginate; *labrum* transverse, obovate, fringed with setæ; *mandibles* broad at apex, tridentate; *cibarial apparatus* short; *tongue* broad, subemarginate and fringed with short hair; *paraglossæ* very slightly longer than the tongue, their apex broadly rounded and fringed with hair; *labial palpi* as long as the tongue, joints subequal, gradating in substance, subclavate; *labium* about as long as the tongue, pyramidal at its apical inosculation; *maxillæ* about as long as the tongue, slightly lanceolate, fringed with short hair;

maxillary palpi rather longer than the maxillæ, with six joints, the basal joint robust and slightly constricted in the middle, the third joint linear and the longest, the remainder gradually decreasing in length and substance.

THORAX subquadrate; *prothorax* transverse, linear, angulated at the sides; *mesothorax* with its *bosses* protuberant; *scutellum* and *post-scutellum* semilunulate; *metathorax* abruptly truncate, and longitudinally carinated in the centre; *wings* with two submarginal cells, a third slightly indicated, the first recurrent nervure springing from the extreme apex of the first submarginal cell, closely to the first transverso-cubital nervure, and the second closely before the termination of the second submarginal cell; *stigma* of the wing large and distinct; *legs* wholly destitute of polliniferous hair, the terminal joint of the tarsus as long as the two preceding; *claws* bifid; ABDOMEN subtruncate at the base, subconical with a downward bias.

The MALE differs in having the *mandibles* distinctly bidentate, the external tooth acute; the *antennæ* are very slightly longer and more curved, and their colouring is more intense and more widely distributed. These insects are glabrous, generally intensely black, dull on the head and thorax, but shining on the abdomen, and are more or less thickly punctured, and they are usually gaily marked with yellow, citron, or red, especially on the face, thorax, and legs.

NATIVE SPECIES.

1. *annulatus*, Fab., ♂ ♀. 2½–3 lines.
 annulatus, Kirby.
2. *dilatata*, Kirby, ♂. 3 lines. (Plate I. fig. 2 ♂.)
 Hylæus dilatatus, Curtis.

3. *annularis,* Kirby, ♂ ♀. 2½–3 lines.
4. *hyalinata,* Smith, ♂ ♀. 2–3 lines.
5. *signata,* Panzer, ♂ ♀. 3–3½ lines. (Plate I. fig. 2 ♀.)
signata, Kirby.
6. *cornuta,* Kirby, ♂ ♀. 3–3¼ lines.
7. *varipes,* Sm., ♂ ♀. 1½ lines.
8. *variegata,* Fab., ♂ ♀. 2–3 lines.

GENERAL OBSERVATIONS.

This genus is named from προσωπίς, apparently in allusion to its seemingly masked face, most of the species having yellow markings more or less conspicuous upon the face.

It is the least pubescent of any of the bees, even less so than those confirmed parasites, the genera *Nomada* and *Stelis,* thus further tending to corroborate its apparently parasitical habits, for none of the truly pollinigerous bees are so destitute of hair. The ground-colour of the species is intensely black, variously decorated on the face, thorax, and legs, with markings of different intensities of yellow; but one of our species, the *P. variegata,* is also gaily marked with red. Indeed exotic species, and especially those of warm climates, are often very gay insects.

They have usually been considered as parasitical insects, from their being unfurnished with the customary apparatus of hair upon the posterior legs, with which pollinigerous insects are generally so amply provided. In contradiction to their parasitism, it is asserted that they have been repeatedly bred from bramble sticks; this circumstance is no proof of the fact of their not being parasitical, for many bees, for instance *Ceratina, Heriades*, etc., nidificate in bramble sticks, and they may have superseded the nidificating bee by depositing their ova in the nests of the latter; although it certainly is a remarkable circumstance that some one of these bees has never escaped destruction in the several instances in which these have been thus bred. It is also said that their nests contain a semi-liquid honey. The fact of the larva of a wild bee being nurtured upon any other provender than a mixture of pollen and honey, does not elsewhere occur, and it would seem to contradict the function this family is ordained to exercise, by conveying pollen from flower to flower, and which besides, in every other case, constitutes the nutritive aliment of the larva. But then, again, the structure of its tongue, which resembles somewhat that of *Colletes* in lateral expansion, and with which it would be provided for some analogous purpose, seems to contradict parasitical habits, although St. Fargeau asserts that it is parasitical upon this genus, and if so, although it has not been observed in this country, the analogous structure of the tongue might be perhaps explained.

But notwithstanding this deficiency of positive characters, from the absence of pollinigerous organs, nature is not to be controlled by laws framed by us upon the imperfect induction of incomplete facts, for if it be incontestable that this genus is constructive and not parasitical, the riddle presented by this structure of its tongue is at once solved, for without any affinity beyond that single peculiarity with *Colletes*, it presents an anomaly of organization which cannot be accounted for but by its application to a use similar to what we find it applied in that extraordinary genus,—a use that could not be extant in a parasite. In *Colletes* it is the concomitant of as ample a power of collecting pollen as any that we find exhibited throughout the whole range of our native bees, but in *Prosopis* it is concurrent with a total deficiency of the ordinary apparatus employed for that purpose.

One of the species of this genus has been found near Bristol, with the indication of a *Stylops* having escaped from it, which is a further extension of the parasitism of that most extraordinary genus, but the *Stylops* frequenting it has not yet been discovered, which would doubtless present a new species, therefore an interesting addition to the series already known.

These insects are not at all uncommon in some of the species during the latter spring and summer months, and they frequent the several *Resedas*, being very fond of Mignonette. They are also found upon the *Dracocephalum Moldavica*, and occur not unfrequently upon the Onion, which in blossom is the resort of many interesting insects. The majority of them emit when captured, and if held within the fingers, a very pungent citron odour, exceedingly refreshing on a hot day, in intense sunshine. Some of the species are rare, especially those very highly coloured, as is also the *P. dilatata*, so named from the peculiar triangular expansion of the basal joint of the antennæ, the female of which is not known or possibly has only been overlooked or not identified. The *P. varipes* and *P. variegata*, which are the most richly coloured, occur in the west of England, and in one, the *P. cornuta*, the clypeus is furnished with a tubercle.

Subsection *b*. LINGUÆ LANCEOLATÆ (with lancet-shaped tongues).

Genus 3. SPHECODES, *Latreille*.

(Plate I. fig. 3 ♂ ♀.)

MELITTA ** *a*, Kirby.

Gen. Char.: HEAD transverse, linear, fully as wide as the thorax, flat, with a slightly convex tendency; *ocelli* in a triangle; *antennæ* short, scarcely geniculated; *face* beneath the insertion of the antennæ, protuberant; *clypeus* transverse, margined, convex; *labrum* transversely ovate, deeply emarginate, in the centre in front; *mandibles* bidentate, obtuse, the external tooth projecting much further than the second; *tongue* short, lanceolate, fringed with setæ; *paraglossæ* not so long as the tongue, abruptly terminated, and setose at the extremity; *labial palpi* not so long as the paraglossæ; the joints comparatively elongate and slender, and decreasing towards the apex in length and substance; *labium* rather longer than the tongue, its inosculation straightly transverse; *maxillæ* about the length of the tongue, broad and lanceolate; *maxillary palpi* six-jointed, the first joint shorter and less robust than the second, which is also shorter and less robust than the third, which is the longest and most robust of all, the terminal joints more slender, and declining gradually in length. THORAX ovate; *prothorax* linear, produced into a sharp tooth on each side; *mesothorax* with longitudinal lateral impressed lines; *bosses* acutely protuberant; *scutellum* quadrate; *postscutellum* inconspicuous; *metathorax* slightly gibbous; *wings* with three submarginal cells, and a fourth slightly commenced, the second narrow, forming a truncated triangle, and receiving the first recurrent nervure in its centre, the second recurrent nervure springing from just beyond the centre of the third submarginal cell; *legs* slightly but rigidly spinose and setose; *claws* bifid. ABDOMEN ovate.

The MALES differ, in having the antennæ longer and sometimes moniliform, the lower part of the *face* and *clypeus* usually covered with a dense short silvery decumbent pubescence, and they have the *metathorax* truncated at its base; in other respects they greatly resemble their females.

The insects of this genus may be called glabrous, their pubescence being so slight and scattered, they usually shine brightly, and are more or less deeply punctured; and the abdomen is always partially or entirely of a bright ferruginous red, sometimes verging into fuscous or pitchy.

NATIVE SPECIES.

1. *gibbus*, Linnæus, ♂ ♀. 3–4½ lines. (Plate I. fig. 3 ♂ ♀).
    sphecoides, Kirby, ♀.
    monilicornis, Kirby, ♂.
    picea, Kirby, ♂.
2. *Geoffroyella*, Kirby, ♂ ♀. 1–3 lines.
    divisa, Kirby, ♂.
3. *fuscipennis*, Germar, ♂ ♀. 4½–6 lines.

GENERAL OBSERVATIONS.

This genus is named from σφὴξ, a wasp, from its apparent resemblance to some of the sand wasps.

They are not uncommon insects, and I have found them abundant in sandy spots sporting in the sunshine upon the bare ground, where they run about with great activity, the females chiefly, the males the while disporting themselves upon any flowers that may be adjacent, and they are especially fond of Ragwort. Their prevalent colours are black and red, the latter occurring only on the abdomen in different degrees of intensity and extension, sometimes occupying the whole of that division of the body, and sometimes limited to a band across it. Much difficulty attaches to the determination of the species from the characters which separate them being extremely obscure, for it is not safe to depend upon the differences of the arrangement of colour upon them, as it varies infinitely; nor can their relative sizes be depended upon as a clue, for in individuals which must be admitted to be of the same species, size takes a wider extent of difference than in almost any of the genera of bees. St. Fargeau, who maintains the parasitism of the genus, accounts for it by saying that in depositing their eggs in the nests of the *Andrenæ, Halicti,* and *Dasypoda,* the *Sphecodes* resorts to the burrows of the species of these genera indifferent to their adaptation to its own size, and thus from the abundance or paucity of food so furnished to its larvæ, does it become a large or a small individual. Westwood says the species are parasitical upon Halictus. Latreille says they are parasites. They are certainly just as destitute of the pollinigerous apparatus as the preceding genus. Mr. Thwaites once thought he had detected a good specific character in the differing lengths of the joints of the antennæ, but I believe he never thoroughly satisfied himself of its being practically available. At all events great difficulty still attaches to their rigid and satisfactory determination. There is an array of entomologists who deny their being parasites. Mr. Kirby says they form their burrows in bare sections of sandbanks exposed to the sun, and nine or ten inches deep, and which they smooth with their tongues. But then, in impeachment of the accuracy of his observation, he further supposes there are three sexes, founding his statement upon what Réaumur remarks of having observed pupæ of three different sizes in the burrows. In the first place, it is not conclusive that these pupæ were those of *Sphecodes*, and secondly we know that this condition of three sexes is found only in the social tribes, wherein the peculiarities of the economy exact a division of offices. Therefore his adoption of this inaccuracy militates against the reception of his other statement. But Smith also states that they are not parasites, and apparently founds his assertion upon direct observation. It still, however, remains a debatable point, from the fact of the destitution of pollinigerous brushes, and thence the character of the food necessary to be stored for the larva. It would be very satisfactory if these apparent inconsistencies could be lucidly explained.

If, however, it be ultimately proved that *Sphecodes* is a constructive bee, as well as *Prosopis*, we have then this fact exhibited by our native genera, that none of the subfamily of our short-tongued bees, or *Andrenidæ*, are parasitical. This is a remarkable peculiarity, as it is amongst them that we should almost exclusively expect to find that distinguishing economy, from the seemingly imperfect apparatus furnished in the short structure of their tongues. It is possible, however, that nature has so moulded them as to fit them chiefly for fulfilling its objects within merely a certain range of the floral reign, and which restricts them to visiting flowers which do not require the protrusion of a long organ to rifle their sweet stores.

Genus 4. ANDRENA, *Fabricius.*
(Plates II. and III.)
MELITTA ** c, Kirby.

Gen. Char.: HEAD transverse, as wide as the thorax; *ocelli* in a triangle on the vertex; *antennæ* filiform, geniculated, the basal joint of the flagellum the longest; *face* flat; *clypeus* convex, transverse, quadrate, slightly rounded in front; *labrum* transverse, oblong; *mandibles* bidentate; *tongue* moderately long, lanceolate, fringed with fine hair; *paraglossæ* half the length of the tongue, abruptly terminated and setose at the extremity; *labium* about half the length of the entire apparatus, its inosculation acute; *labial palpi* inserted above it, below the origin of the paraglossæ in a sinus upon the sides of the tongue; *maxillæ* irregularly lanceolate; *maxillary palpi* six-jointed, longer than the maxillæ, the basal joint about as long as the fourth, but more robust, the second joint the longest, the rest declining in length and substance. THORAX ovate; *prothorax* not distinct; *mesothorax* quadrate; *bosses* protuberant; *scutellum* lunate; *post-scutellum* lunulate; *metathorax* gibbous, and pubescent laterally; *wings* with three submarginal cells, and a fourth slightly commenced, the second quadrate, and with the third receiving a recurrent nervure about their middle; *legs* densely pubescent, especially externally, and particularly the posterior pair, which have a long curled lock upon the trochanter beneath, the anterior upper surface of the femora clothed with long loose hair, which equally surrounds the whole of the tibiæ, but which is less long upon their plantæ, the *claws* strongly bifid. ABDOMEN ovate, a dense fringe edging the fifth segment, and the terminal segment having a triangular central plate, its sides rigidly setose.

The MALE differs in having the *head* rather wider than the thorax, the *vertex* where the ocelli are placed more protuberant, the *mandibles* very large and more acutely bidentate, sometimes largely forcipate and with but one acute tooth; the males in most species greatly differ from their females.

None of these insects exhibit any positive colouring of the integument, excepting in some upon the abdomen, which exhibits red bands, and is disposed to vary considerably in intensity and breadth, and in some the *clypeus* and *face* are of a cream-colour, but which occurs chiefly among the males. They are very dissimilar in general appearance, some being densely pubescent all over, others merely so on the head and thorax; others are banded with white decumbent down, and some are wholly unmarked upon the abdomen. These peculiarities help to group them, and thus facilitate their recognition.

NATIVE SPECIES.

§ *Banded with red on the abdomen, the segments of which are more or less fringed.*

1. *Hattorfiana*, Fab., ♂ ♀. 6–7 lines.
   Lathamana, Kirby, ♀.
   hæmorrhoidalis, Kirby, ♀.
2. *zonalis*, Kirby, ♂ ♀. 4½–5 lines.
3. *florea*, Fabricius, ♂ ♀. 5–6½ lines.
   Rosæ, Kirby, var.
4. *Rosæ*, Panzer, ♂ ♀. 4–6 lines. (Plate III. fig. 1 ♂ ♀.)
   Rosæ, Kirby, ♀.

5. *decorata*, Smith, ♂ ♀. 5–6½ lines.
6. *Schrankella*, Kirby, ♂ ♀. 4–5 lines.
   affinis, Kirby.
7. *cingulata*, Fabricius, ♂ ♀. 3½–4 lines. (Plate III. fig. 3 ♂ ♀.)
   cingulata, Kirby.

§§ *Abdominal segments edged with decumbent short down, or fringed with long hair.*

8. *longipes*, Shuckard, ♂ ♀. 4–6 lines. (Plate III. fig. 2 ♂ ♀.)
9. *chrysosceles*, Kirby, ♂ ♀. 3½–4½ lines.
10. *dorsata*, Kirby, ♂ ♀. 4–4½ lines.
    combinata, Kirby.
    nudiuscula, Kirby.
11. *connectens*, Kirby. 5 lines.
12. *Wilkella*, Kirby, ♀. 5¾ lines.
13. *Coitana*, Kirby, ♂ ♀. 4 lines.
    Shawella, Kirby.
14. *labialis*, Kirby, ♂ ♀. 5½–6 lines.
15. *Lewinella*, ♂. 3¾ lines.
16. *xanthura*, Kirby, ♂ ♀. 3½–6 lines.
    ovatula, Kirby.
17. *Collinsonana*, Kirby, ♂ ♀. 3½–4½ lines.
    digitalis, Kirby.
    proxima, Kirby.
18. *albicrus*, Kirby, ♂ ♀. 4–5½ lines.
    barbilabris, Kirby.
19. *minutula*, Kirby, ♂ ♀. 2½–3½ lines.
    parvula, Kirby.
20. *nana*, Kirby, ♀. 3½ lines.
21. *convexiuscula*, Kirby, ♂ ♀. 5 lines.
22. *Kirbyi*, Curtis, ♀. 6 lines.

23. *fuscata*, Kirby, ♀. 4½ lines.
24. *Afzeliella*, Kirby, ♂ ♀. 4½–5 lines.
25. *fulvicrus*, Kirby, ♂ ♀. 3½–5¼ lines.
    contigua, Kirby.
26. *fulvago*, Christ. ♂ ♀. 4–4½ lines.
    fulvago, Kirby.
27. *tibialis*, Kirby. 5–7¼ lines.
    atriceps, Kirby.
28. *Mouffetella*, Kirby, ♂ ♀. 5–7 lines.
29. *nigro-ænea*, Kirby, ♂ ♀. 5–6½ lines.
30. *bimaculata*, Kirby, ♂. 5½ lines.
31. *Trimmerana*, Kirby, ♂ ♀. 5–6 lines.
32. *conjuncta*, Smith, ♀. 5½ lines.
33. *varians*, Rossi, ♂ ♀. 4–5½ lines.
34. *helvola*, Linnæus, ♂ ♀. 5–5½ lines.
    picipes, Kirby, ♂.
    angulosa, Kirby.
35. *Gwynana*, Kirby, ♂ ♀. 4–5½ lines.
    pilosula, Kirby.
36. *angustior*, Kirby, ♂ ♀. 4–5 lines.
37. *picicornis*, Kirby, ♂ ♀. 5–6 lines.
38. *spinigera*, Kirby, ♂ ♀. 5–6 lines.
39. *Smithella*, Kirby, ♂ ♀. 3–6 lines.
40. *Lapponica*, Zetterstedt, ♂ ♀. 3½–5½ lines.
41. *tridentata*, Kirby, ♂. 4½ lines.
42. *denticulata*, Kirby, ♂ ♀. 4–5½ lines.
    Listerella, Kirby.
43. *nigriceps*, Kirby, ♀. 5 lines.
44. *pubescens*, Kirby, ♂ ♀. 4–5 lines.
    rufitarsis, Kirby.
    fuscipes, Kirby.

§§§ *Thorax very pubescent, abdomen smooth and shining.*

45. *albicans*, Kirby, ♂ ♀. 4–5 lines.
46. *pilipes*, Fabricius, ♂ ♀. 5–7 lines.
    pratensis, Kirby.
47. *cineraria*, Linnæus, ♂ ♀. 5–7 lines. (Plate II. fig. 2 ♂ ♀.)
    cineraria, Kirby.
48. *thoracica*, Fabricius, ♂ ♀. 5–7½ lines.
    thoracica, Kirby.
    melanocephala, Kirby.
49. *nitida*, Fourcroy, ♂ ♀. 5–6½ lines. (Plate II. fig. 3 ♂ ♀.)
    nitida, Kirby.
50. *vitrea*, Smith, ♀. 6½ lines.

§§§§ *The entire body densely pubescent.*

51. *fulva*, Schrank, ♂ ♀. 4–6½ lines. (Plate II. fig. 1 ♂ ♀.)
    fulva, Kirby.
52. *Clarkella*, Kirby, ♂ ♀. 4½–6½ lines.

GENERAL OBSERVATIONS.

Fabricius seems to have named this genus from ανθρήνη, *a wasp*, but why, it is impossible to say. Although one name is as good as another, it being indifferent what the name may be, yet where so evident an attempt to give a name pertinence is conspicuous, it is remarkable that it should be so little relevant, for none of the characteristics of a wasp or hornet are exhibited in these insects.

Possibly it was from the genus being the most numerous in species that Dr. Leach was induced to give

this subfamily its collective designation, making the
other genera thus converge to it as to a centre. He
took its elliptical form as typical. Indeed, it is remarkable how very judiciously this was done, for it is a form
not apparent among the normal bees excepting in two
exceptional cases, the one upon the frontiers of this
subfamily, in almost debatable land, where the last of
the *Andrenidæ* and the first of the *Apidæ* seem almost
to melt into one another; and in the other case, in the
parasitical *Nomada*, whose parasitism is in every instance, but one only, restricted to the first subfamily.
A different type of form prevails amongst the *Apidæ*,
upon which I shall have subsequently occasion to speak.

These insects are not distinguished for any elaborate
economy. Varying in the species, some prefer vertical
banks, others sloping undulations, and again others horizontal flat ground or hard down-trodden pathways.
Some burrow singly, and others are gregarious, collected in great numbers upon one spot. They are,
perhaps, the most inartificial burrowers of all the bees.
Their tunnels vary from five to nine or ten inches in
depth, and in some species they are formed with other
small tunnels slanting off from the main cylinder. The
sides and bottom are merely smoothed, without either
drapery or polish. The little cells thus formed are then
supplied with the usual mixture of pollen and honey
kneaded together, which in the larger species forms a
mass of about the size of a moderate red currant, its
instinct teaching it the quantity necessary for the nurture of the young which shall proceed from the egg
that it then deposits upon this collected mass of food.
The aperture of each little tunnel is closed with particles of the earth or sand wherein the insect burrows,

and it proceeds to the elaboration of another receptacle
for a fresh brood until its stock of eggs becomes exhausted. Some species have two broods hatched in the
year, especially the earlier ones,—for several present
themselves with the earliest flowers,—but others are restricted to but one. The quantity of pollen they collect is considerable, and in fact they are supplied with
an apparatus additional to what is furnished to any of
the other genera in a curled rather long lock of hair
that emanates from the posterior trochanters. This, with
the fringes that edge the lower portion and sides of the
metathorax, as well as the usual apparatus upon the
posterior legs, enables the insect to carry in each flight
home a comparatively large quantity of pollen, but perhaps scarcely enough at once for the nurture of one
young one, and it therefore repeats the same operation
until sufficient is accumulated.

The exact period occupied by their transformations is
not strictly known; it will, of course, vary in the species, as also in those in which two broods succeed each
other in the year, but the larva rapidly consumes its
store and then undergoes its transformation. It does
not spin a cocoon, but in its pupa state it is covered
all over with a thin pellicle, which adheres closely to
all the distinct parts of the body. It is not known how
this is formed; perhaps it is a membrane which trans-
udes in a secretion through the skin of the larva, or it
may be this itself converted to its new use, which seems
to be for the protection of all the parts of the now
transmuting imago, until these in due course shall have
acquired their proper consistency.

These insects in their perfect state vary very considerably in size, both individually and specifically, the

former depending upon both the quantity and quality of the food stored up, for the pollen of different plants varies possibly in its amount of nutriment, else why should we observe so marked a difference in the sizes of individuals whose parent instinct would prompt to furnish them with an uniform and equal supply. The differences of specific appearance is often very considerable in long genera, and perhaps in no genus is it more conspicuously so than in *Andrena*, for here we have some wholly covered with dense hair, and others almost glabrous; others again with the thorax only pubescent; some are black, some white, some fulvous, or golden tinted, and some red; some we find banded with decumbent down, and others with merely lateral spots of this close hair, but the most prevalent colour is brown, which will sometimes by immaturity take a fulvous or reddish hue. In many males we see excentrically large transversely square heads broader than the thorax, which also have widely spreading forcipate mandibles, with often a downward projecting spine at their base beneath; and it is chiefly these extravagantly formed males which are most dissimilar to their own partners that the result of observation alone confirms their specific identity. In other cases the males are so like their females that a mere neophyte would unite them. In many males the clypeus and labrum are white, which also occurs in some females; for instance, in *A. labialis*, but this peculiarity is found more rarely in this sex. The species are much exposed to the restricting influences of several parasites, whose parasitism is of a varying character, but the term should properly be applied only to the bees which deposit their eggs in their nests, and whose young, like that of the cuckoo among the birds, thrives at the expense of the young of the sitos by consuming its food, and thus starving it. These parasites consist of many of the species of *Nomada*, very pretty and gay insects, but in every case totally unlike the bee whose nest they usurp. Several of the species of these *Nomadæ* are not limited to any particular species of *Andrena*, but infest several indifferently, whereas others have no wider range in their spoliation than one single species, to which they always confine themselves. In my observations under the genus *Nomada* I shall notify those which they assail amongst the *Andrenæ*, as well as the other genera which they also infest.

The others which attack them are more properly positive enemies than parasites, for they prey upon the bees themselves, or, as in the case of the remarkable genus *Stylops*, render the bee abortive by consuming its viscera and ovaries. I have spoken of these insects in the chapter upon parasites, to which I must refer, but I may here add that the female is apterous, and never quits the body of the bee. Much mystery attaches to their history in which their impregnation is involved, for the male, immediately upon undergoing its change into the imago, escapes through the dorsal plates of the abdomen of the bee wherein it was bred and takes flight. In localities where they occur they may be usually taken on the wing in the month of May. The female would seem to be viviparous, and produces extraordinary multitudes at one birth, extending to hundreds. Being born as larvæ within the body of the bee they seek to escape from their confinement, and find the opportunity in the suture which separates the mesothorax from the metathorax. Their extreme minuteness admits of their passing through the very constricted tube which connects the abdomen with the thorax. Having now escaped into the air they alight upon the flowers which the bee frequents, and thence they affix themselves to other bees which may visit these plants, and thus perpetuate the activity of the function it is their instinct to fulfil. That many may be lost there can be no question; but Nature is very prodigal of life, for by life it endows life, and thus its activity is enlarged to a wider circle. Although the matured *Stylops* has preyed upon all the internal organs of the bee its attack is not immediately fatal, although the life of the creature may be thus considerably abridged, but it seems to live sufficiently long afterwards to disseminate the distribution of the *Stylops*. A small blackish *Pediculus*, which Mr. Kirby called *Pediculus Melittæ*, is found also both upon the flowers the bees frequent and also upon the bees themselves, especially the pubescent ones; but this insect is not limited to the genus *Andrena*, as I shall have occasion to notice. The flower I have chiefly found them upon is the Dandelion (*Leontodon*). Their peculiar economy and connection with the bees is unknown; it may be merely an accidental and temporary attachment, but they even accompany them to their burrows.

Another and more curious case of attack upon the young of the *Andrena*, is instanced in the reputed parasitism of the Coleopterous genus *Meloë*. The perfect insect is a large apterous, fleshy, heteromerous beetle, ten times as big as the bee. Its vermicle, having issued from the egg, has the appearance of a very small pediculus, of an orange colour. They are often seen upon flowers, and, like the former pediculus, attach themselves to such suit-

P

able *Andrena* as may happen to visit the flowers they are upon; and, it is said, that they are thus conveyed by the bee to its domicile, and there feed to maturity upon the larva of the bee. I have no faith in the correctness of this statement, for it is not credible that so small a creature as the larva of an *Andrena* could fully feed the larva of so large a beetle. Observation has not satisfactorily confirmed it, and the connection may be, as in the former case, merely accidental.

Although, perhaps, not a strictly scientific course, it is certainly a matter of convenience in very long genera to break them up into divisions, framed upon external characters, readily perceptible, and, by which means, the species sought for may be more readily found. This I have done in the preceding list of the species, and which are based upon very prominent features. A slight divarication from the typical neuration of the wing is observed in some species, but it is not of a sufficiently marked character to afford a divisional separation, and even much less a subgeneric one. I have therefore passed it unnoticed. The commencing entomologist will often find considerable difficulty at first in determining the species of this genus, for so much depends upon condition; and where the colour of the pubescence is the chief characteristic, a very little exposure to the atmosphere much alters their physiognomy, but time, patience, and perseverance will ripen the novice into an adept. The connection of the males with the females, from their ordinarily great dissimilarity, was only to be accomplished by positive observation, but now that this, in the majority of cases, is effected, good descriptions facilitate their discrimination.

The most conspicuous species are the *Hattorfiana* and

the *Rosæ* for size and colour; the *Schrankella* is also a very pretty species; and perhaps the commonest of all the *cingulata* is the prettiest of all, with its yellow nose and red abdomen; in the next section we may point out the *longipes* as being a very elegant insect,* as are also the *chrysosceles* and the *helvola*. In this section we find those most subject to the attacks of the *Stylops*, for instance the *labialis, convexiuscula, picicornis, Afzeliella, nigro-ænea, Trimmerana, Gwynana*, etc. The whole of the third and fourth sections are splendid insects, especially the *fulva* in the last. The comparative rarity of some results chiefly from an exceedingly local habitat. Many of the species may be found everywhere where insects can be collected, consequently, all over the United Kingdom. In all the three seasons of the year, which prompt animal life, some of the species may be collected, and the flowers they chiefly prefer are the catkins, especially of the sallow, the early flowering-fruits, the hedgerow blossoms, the heath, the broom, the dandelion, chickweed, and very many others.

Genus 5. CILISSA, *Leach.*

(Plate V. fig. 1 ♂ ♀.)

MELITTA ** *c*, partly, Kirby.—ANDRENA, Fab. Latreille.

Gen. Char.: HEAD transverse, scarcely so wide as the thorax, flat; *ocelli* in an open triangle on the vertex;

---

* This insect was first captured by me, and with this, my manuscript name, attached to it, it was distributed to entomologists with an unsparing hand. The ordinary courtesy of the science has been, for the describer, when not the capturer, to adopt and circulate the original authority, and not to appropriate it. Similar buccaneering has been

face flat; *clypeus* transverse, margined; *labrum* transverse, slightly rounded in front; *mandibles* bidentate; *cibarial apparatus* moderately long; *tongue* lanceolate, fringed with delicate hair; *paraglossæ* about one-third the length of the tongue, abruptly terminated, lacerate and setose at the extremity; *labial palpi* rather longer than the paraglossæ, the basal joint considerably the longest, all the joints subclavate and diminishing both in robustness and length to the apex; *labrum* half the length of the entire apparatus, its inosculation acutely triangular; *maxillæ* subhastate, as long as the tongue; *maxillary palpi* six-jointed, less than half the length of the maxillæ, the joints short, subclavate and decreasing gradually from the base to the apex. THORAX densely pubescent, obscuring its divisions; *metathorax* truncated; *wings* with three submarginal cells, and a fourth slightly commenced, the second subquadrate and receiving the first recurrent nervure in its centre, the second recurrent nervure issuing from beyond the centre of the third submarginal cell; *legs* all pilose, especially the posterior pair, which have hair beneath the *coxæ* and *trochanters*, above only on their femoræ, but surrounding the *tibiæ*, and as dense externally upon their *plantæ*; *claws* distinctly bifid. ABDOMEN ovate, truncated at the base, the segments banded at their apex, with decumbent down, which becomes densely and widely setose on the fifth segment, the terminal segment having a central triangular glabrous plate, carinated down the centre, and very rigidly setose laterally.

The MALE scarcely differs, except in having the *antennæ*

practised with poor Bainbridge's *Osmia pilicornis*, to which he had attached this manuscript name, he being the first to introduce it, having caught it at Birchwood.

less distinctly geniculated, the flagellum taking a sweeping curve, the *face* and *clypeus* much more pubescent, but the *legs* sexually less so; the sexes are much alike.

NATIVE SPECIES.

1. *tricincta*, Kirby, ♂ ♀. 5 lines. (Plate V. fig. 1 ♂ ♀.)
  ? *Apis leporina*, Panzer.
2. *hæmorrhoidalis*, Fab. ♂ ♀.
  hæmorrhoidalis chrysura, Kirby.

GENERAL OBSERVATIONS.

This genus has been named without any reference to any peculiarity, Dr. Leach having applied a Proper name to it to designate it.

The *Cilissa tricincta* is perhaps most like the larger species of the genus *Colletes*, both in markings and in the form of the body, but in resemblance of form the second species participates. Although robust insects, and as large as the larger *Andrenæ*, they are yet unprovided with the same ample means for conveying pollen, being destitute of the lock of hair upon the posterior trochanters and the sides of the metathorax are less densely pubescent. The ground colour is brown. Their economy is assumed to resemble that of *Andrena*, although it has not been so closely investigated; for my own part I have never had the opportunity of tracing it to its nidus, having always captured the species upon flowers. They are fond of the trefoil (*Trifolium repens*), and the *C. chrysura* frequents the *Campanula rotundifolia*, as well as the flowers of the throatwort (*Trachelium*). In their excursions they are usually accompanied by their males. Both species are found in the south and west of England.

Section 2. *With entire paraglossæ.*

Subsection c. LINGUÆ ACUTÆ (acute tongues).

a. *With three submarginal cells to the wings.*

Genus 6. HALICTUS, *Latreille.*
(Plate IV.)

MELITTA ** *b*, Kirby.

Gen. Char.: HEAD transverse, flattish, scarcely so wide as the thorax; *ocelli* in an open triangle on the vertex, which is flat; *antennæ* short, filiform, geniculated, scape quite or more than half as long as the flagellum; *face* flat, excepting in the centre just below the insertion of the antennæ, where it is protuberant; *clypeus* transversely lunulate, very convex; *labrum* subquadrate, very convex, with a central, linear, carinated appendage in front, nearly as long as the basal portion; *cibarial apparatus* moderate; *tongue* very acute and delicately fringed with short hair; *paraglossæ* acute, about half the length of the tongue; *labial palpi* not quite so long as the paraglossæ, the basal joint very long, the rest decreasing gradually in length; *labium* about as long as the tongue, its inosculation emarginate; *maxillæ* subhastate, rather longer than the tongue; *maxillary palpi* filiform, the basal joint the shortest, second the longest, the rest decreasing in length. THORAX oval, usually pubescent, sometimes glabrous; *prothorax* inconspicuous, as are the *bosses* of the mesothorax; *scutellum* and *post-scutellum* lunulate, the former convex; *metathorax* gibbous or truncated, but laterally pubescent even in the glabrous species; *wings* with three submarginal cells, and a fourth sometimes commenced, the second subquadrate and receiving the first recurrent nervure close to its extremity, the second being received beyond the centre of the third submarginal cell [a slightly different arrangement takes place in some of the species, which will be noticed subsequently] ; the *legs* all setose, but the setæ not very long, and the posterior *coxæ* and *trochanters* have long hair beneath; the *claws* bifid. ABDOMEN ovate, the terminal segment with a longitudinal linear incision in its centre.

The MALES differ in having the antennæ as long or longer than the thorax; the *labrum* transverse, linear, and the *abdomen* usually elongate and cylindrical, and much longer than the head and thorax.

NATIVE SPECIES.

1. *xanthopus*, Kirby, ♂ ♀. 4–5½ lines. (Plate IV. fig. 1 ♂ ♀.)
  Lasioglossum tricingulum, Curtis.
2. *quadricinctus*, Fabricius, ♂ ♀. 4–4½ lines.
  quadricinctus, Kirby.
3. *rubicundus*, Christ. ♂ ♀. 4–5 lines.
  rubicundus, Kirby.
4. *cylindricus*, Fabricius, ♂ ♀. 3–5 lines.
  malachura, Kirby.
  fulvo-cincta, Kirby.
  abdominalis, Kirby.
5. *albipes*, Fabricius, ♂ ♀. 3–4 lines.
  albipes, Kirby.
  obovata, Kirby.
6. *lævigatus*, Kirby, ♂ ♀. 3–4½ lines.
  lugubris, Kirby.
7. *leucozonius*, Schrank, ♂. 3–4½ lines.
  leucozonius, Kirby.
8. *quadrinotatus*, Kirby, ♂ ♀. 2–3 lines.
9. *sexnotatus*, Kirby, ♂ ♀.

10. *lævis*, Kirby, ♀. 4 lines.
11. *fulvicornis*, Kirby, ♂. 4 lines.
12. *minutus*, Kirby, ♂ ♀. 2½-3½ lines.
13. *nitidiusculus*, Kirby, ♂ ♀. 2-3 lines.
14. *minutissimus*, Kirby, ♂ ♀. 1½-2½ lines. (Plate IV. fig. 3 ♂ ♀.)
15. *flavipes*, Kirby, ♂ ♀. 3-4 lines. (Plate IV. fig. 2 ♂ ♀.)
seladonia, Kirby.
16. *Smeathmanellus*, Kirby, ♂ ♀. 2½-3½ lines.
17. *æratus*, Kirby, ♂ ♀. 2½-3 lines.
18. *leucopus*, Kirby, ♂ ♀. 3-3½ lines.
19. *morio*, Kirby, ♂ ♀. 2-2½ lines.

GENERAL OBSERVATIONS.

This genus was named by Latreille from ἁλίζω, *to crowd*, or *collect together*, from the fact of their nidificating in numbers on the same spot.

The females closely resemble in form those of the genus *Andrena*, but the males are very unlike both those of that genus and their own females, for they all have long cylindrical bodies and very long antennæ, much longer relatively than those of the former genus. Although none of the species approach in size the larger ones of the preceding genus, their extremes of specific size are as distant apart as they are in that genus, the smallest being extremely minute. Some of even the commoner species are very pretty when in fine condition, and several of them have a rich metallic green or blue tint, and in the majority the wings are iridescent with the brightest and gayest colours of the rainbow. The numbers in which they associate together upon the same spot varies considerably, and a very few indeed burrow solitarily and apart from their congeners. In burrowing they form a tunnel which branches off to several cells, the excavations being as inartificial as are those of *Andrena*. Walkenaer tells us in his memoir upon the genus *Halictus*, that they line their cells with a kind of glaze, that they burrow in horizontal surfaces to a depth of about five inches, and which they polish very smoothly previous to covering it with their viscous secretion, and that the cells are all oval, the largest end being at the bottom. He says also that they burrow solely during the night, especially when the moon is shining, when it is difficult to walk without treading upon them; so numerous are they, indeed, that they look like a cloud floating close to the surface of the ground. Although burrowing thus at night, it is only during the day that they supply their nests with their provision of pollen and lay their eggs. Each of their cells is furnished with a small ball of pollen, varying in size with the species, but which never entirely fills the cell, and is affixed intermediately between both extremities, and upon the mass contained in each cell they deposit their small egg, which is placed at the extremity of the lump of pollen most distant from the entrance. The larva is hatched in about ten days, when it changes into the pupa. Some doubt attaches as to the length of time that the pupa remains before its transformation into the imago, and also as to the period at which this takes place. A peculiarity attends the appearance of the larger species. Some are very early spring insects, among which is the *Halictus rubicundus;* this I have seen in abundance on the first fine spring days collecting its stores on the flowers of the chickweed. It is then in the very finest condition, and it is really a very beautiful although a very common insect, having a richly golden fulvous pubescence on the thorax, an intensely black and glabrous abdomen, the apex of which is fringed with golden hair. No males are now to be found at all. Yet it is only some species, and these the larger ones, which are subject to this peculiarity, for the smaller ones I have found burrowing during the summer months in vertical or sloping banks with a sunny aspect, whilst the males were hovering about both in the vicinity and close by, sometimes either playing or fighting on the wing with the very small *Nomadæ*, which infest these species parasitically, whilst their females were sedulously pursuing their vocation. Gradually these joyous spring insects lose their gayness and their brilliancy, as do those which have followed in succession of development with the growing year, and they become senile and faded and are lost as they have progressively fulfilled their function. By this time the ragwort is in bloom, and the thistle displays its pinky blossoms; now the males are to be found numerously exhibiting themselves upon these flowers, and also another equally fresh brood to those of the spring and early summer, of females. My friend the late Mr. Pickering, who was in the early days of the present Entomological Society, when it held its meetings in Old Bond Street, its honorary curator, and who was then and always, even when less leisure was afforded him from professional duties, a most assiduous and diligent observer of the habits of insects, propounded his theory, both in conversation and before the meetings of the Society, although he never drew up a paper upon the subject, that these females were then impregnated, upon which they retired to a hibernaculum, and there remained until the breath of a new spring brought them forth in all the beauty of their gay attire, and that it was from their broods deposited thus in the spring and early summer, that the autumnal insects were developed. This theory is both plausible and possible, and I have no doubt that it is the correct one; and thus is explained the total absence of males at the time of the appearance of the females in the foremost portions of the year; this habit we shall find also in the *Bombi.*

The flowers they delight in, besides those previously named, are among others the ribwort plantain, and the bramble, as well as the *Umbelliferæ* and the flowers of the broom. The females possess two remarkable distinctions of structure not found in any of the other bees, which consist in an articulated appendage in the centre of the front margin of the labrum, and a vertical cleft in the terminal segment of the abdomen, both of which will necessarily have their uses in the economy of the insect, although what these may be has not been discovered.

They, like *Andrena,* are exposed to parasites and enemies. The smaller species of *Nomada* infest their smaller kinds, and St. Fargeau tells us that the *Sphecodes* are also parasitical upon them. The smallest of the genus, which is indeed an exceedingly minute insect, is subject to a very minute strepsipterous destroyer; whether this be a genuine *Stylops* I am not aware, but the supporting insect being so minute, in fact the smallest of our bees, how small must be the enemy bred within it! Another genus of this order has been found by Mr. Dale upon them, and which is figured as the genus *Elenchus* in Curtis's 'British Entomology.' The smaller species are also attacked, upon their return home laden, by spiders and ants. *Chryses* and *Hedychra* are bred at their expense, and some of the *Ichneumons* attack them, as well as the fossorial *Hymenoptera* of the genera *Cerceris*, *Crabro*, and *Philanthus*, and these latter carry them off bodily to furnish their own nests with pabulum. Several of the species exhale a rich balmy odour, and, like all the *Andrenidæ*, they are silent on the wing, and their sting is innocuous and not painful. The males are very eager in their amours, and are not easily repulsed.

Some of the species vary slightly in the neuration of the wings, and this being a rather numerous genus, although not nearly approaching the extent of *Andrena*, it has been proposed to make use of it for its division, but I think this is scarcely required, it not being sufficiently abundant to cause any inconvenience, the species being so distinctly marked in their specific differences by the aid of the metallic brilliancy of several of them. I have therefore arranged the species in the above list in connective order without intermission, and have placed in juxtaposition those species which appear the closest in affinity.

---

b. *With two submarginal cells to the wings.*

Genus 7. MACROPIS, *Panzer*.

(Plate V. fig. 2 ♂ ♀.)

Gen. Char.: HEAD transverse, as wide as the thorax, flattish; *ocelli* placed in a very open curve upon the vertex; *face* flat, but convex in the centre beneath the insertion of the antennæ; *clypeus* very slightly convex; *labrum* transverse, narrowly lunulate; *mandibles* bidentate; *cibarial apparatus* moderately long; *tongue* very acute and fringed with delicate down; *paraglossæ* barely half the length of the tongue, and acute, their apex fringed laterally with down; *labial palpi* inserted in a deep sinus, filiform, the basal joint the longest, the rest diminishing both in length and substance; *labium* about half the length of the entire organ, its inosculation emarginate; *maxillæ* hastate, rather longer than the tongue; *maxillary palpi* six-jointed, the basal joint the shortest, the third the longest, the remainder diminishing gradually in length, and all declining in substance from the basal joint. THORAX oval, rather pubescent; *prothorax* transverse, curving to the mesothorax, whose *bosses* are inconspicuous; *scutellum* transverso-quadrate; *post-scutellum* transverse linear; *metathorax* truncated. WINGS with two submarginal cells, and a third commenced, the second about as long as the first, and receiving both the recurrent nervures, the first near its commencement, and the second nearer its extremity; *legs* robust, with the posterior *tibiæ* and *plantæ* densely clothed externally with short hair; the *plantæ* broad; the second joint of the *tarsus* inserted at the lower angle of the *plantæ*; *claws* bifid. ABDOMEN subtriangular, truncated at its base, not longer than the thorax.

The MALE differs in having the *antennæ* as long as the thorax and curved; the *posterior coxæ* very large and robust, the *trochanters* small and triangular; the *femora* large and much swollen in the centre, the posterior *tibiæ* very large and triangular and convex externally, and the *plantæ* longer than the rest of the tarsus, and slightly curved beneath longitudinally.

NATIVE SPECIES.

1. *labiata*, Panzer, ♂ ♀. 4–4½ lines.
(Plate V. fig. 2 ♂ ♀.)

GENERAL OBSERVATIONS.

The name of this genus comes from μακρὸς, *long*, and ὤψ, *face*, in allusion to the length of that portion of the head, although this assumed discriminative characteristic is scarcely suitable; this again constitutes another of the many instances wherein it would have been much preferable to have imposed a name without any significancy than one which is not thoroughly applicable. It is, indeed, always dangerous to attach a name to a new genus which has reference to some individual peculiarity, for it may eventually exhibit itself as limited to the one single species or sex to which it was originally applied, as to every other subsequently discovered species in the genus it may be inappropriate.

Nothing, so far as I am aware, is known of the habits of these singular insects, which, I believe, have been caught only three times in this country and then only the male sex.

The first, which is in the collection of the British Museum, was brought by Dr. Leach from Devonshire; the second was caught in the New Forest by the late John Walton, Esq., distinguished for his knowledge of the British *Curculionidæ*, and who kindly presented it to me for my collection when I was at the zenith of my enthusiasm for the Hymenoptera, and with that collection it passed to Mr. Thomas Desvignes, in whose possession it remains; and the third was caught by Mr. Stevens, at Weybridge, in Surrey. Why I enter so particularly into these circumstances is, that the genus is extremely peculiar both for scientific position and for structure. In the latter the male is extremely like the male of *Saropoda* and its female is more like the female *Scopulipedes* among the *Apidæ* than one of the *An-*

drenidæ, especially in the form of the abdomen and of the intermediate and posterior legs, as well as in the length of the claws and the low insertion of the posterior joints of the tarsi upon their plantæ, a peculiarity not occurring in another genus of the *Andrenidæ*.

I have no doubt, also, that they are very musical in their flight and are, perhaps, as shrill-winged as is *Saropoda*; whereas one of the great characteristic specialities of the *Andrenidæ* is their silence. This genus, although restrained within the circuit of the subnormal bees by the structure and folding of its tongue, has so much of the habit of one of the true *Apidæ* that it almost prompts the wish to resuscitate the circular systems and place it within its own circle in analogical juxtaposition to *Saropoda* in the circle of the *Apidæ*, where they might impinge one upon the other. It is not often that so rare an insect is at the same time so curious and so suggestive. Having been found, there is no reason why it may not be again found with due and patient diligence; my own experience has taught me how easy it is even in well-hunted ground to make rarities common, within almost a stone's throw of the metropolis, at Hampstead, Highgate, and Battersea, from which localities in the course of my entomological career I have introduced to our fauna many novelties, one of which was certainly a remarkable discovery, from the last spot named, which it is worth recording. A quantity of soil had been removed from the City where an artesian well was being bored, and consequently from varying depths, and carted thence and cast upon the edge of the river-bank at Battersea. The following season, from this soil, a thick and prodigious quantity of the common mustard plant shot up, and when in flower

I happened to be collecting near the spot on the day of our gracious Queen's coronation, when I captured multitudes of a splendid large *Allantus*, entirely new to the British fauna, and a choice addition to collections. This ground had been hunted at all seasons through all botanical and entomological time, and neither had the mustard plant been found there before nor had the insect. Whence did they both come? These observations have certainly nothing to do with the subject in hand, beyond suggesting that with untiring energy in the vicinities indicated where *Macropis* has been already found it may possibly turn up in abundance.

Genus 8. DASYPODA, *Latreille*.

MELITTA ** c, partly, Kirby.

(Plate V. fig. 3 ♂ ♀.)

Gen. Char.: HEAD transverse; *vertex* glabrous; *ocelli* placed in a curved line; *antennæ* short, filiform, geniculated, the scape thickly bearded with long hair and scarcely half the length of the flagellum; *face* and *clypeus* densely pubescent, the latter slightly convex; *labrum* transverse, linear, slightly rounded in front; *mandibles* arcuate, bidentate, the teeth acute and robust; *cibarial apparatus* moderately long; *tongue* long, very acute, and fringed with delicate hair; *paraglossæ* about one-third the length of the tongue, very slender, and acute; the *labial palpi* inserted upon the junction of the labium, very slender, filiform, of uniform thickness, the joints subclavate, the basal joint considerably the longest, the second joint also long, the two terminal joints much shorter and decreasing in length; *labium* about the length of the tongue, its inosculation acutely triangular; *maxillæ* hastate, as long as the tongue; *maxillary palpi* six-jointed, rather more than half the length of the maxillæ, slender, the basal joint the most robust, the second the longest, the rest declining both in thickness and length. THORAX oval, densely pubescent, the divisions indistinct from its density; *scutellum* lunulate; *metathorax* subtruncate; *wings* with two submarginal cells and a third commenced, the second receiving both the recurrent nervures, the first close to its commencement and the second just beyond its centre; *legs* slender, pubescent, especially the *tibiæ* and *plantæ*, the hair upon the posterior pair being extremely dense and long, and each hair twisted minutely spirally; their *coxæ, trochanters*, and *femora* also covered with long hair; *claws* bifid, the inner tooth very short. ABDOMEN oval, the basal and fifth segments densely hairy, the superior surface glabrous and shining, excepting where the white decumbent bands broadly edge the three intermediate segments.

The MALE differs in being more densely pubescent, especially upon the abdomen, which is not glabrous, and in not having the *antennæ* geniculated; the bands of the *abdomen* are fulvous, and its legs are longer and more slender, and it is sexually less hairy, although still considerably so.

NATIVE SPECIES.

1. *hirtipes*, Fab., ♂ ♀. 6–7 lines. (Plate V. fig. 3 ♂ ♀.)
Swammerdamella, Kirby.

GENERAL OBSERVATIONS.

This genus is named from the extreme hairiness of its posterior legs, δασὺς, *hairy*, ποῦς, ποδὸς, *foot or leg*.

It is one of the most elegant of our native bees, both in form and the extreme congruity of its habiliment. This is unfortunately but a bridal raiment, for almost as soon as the arduous duties of maternity supervene these bright garments fade, and the workday suit immediately shows the wear and tear produced by the labours of life. The male flaunts about longer in the freshness of his attire, but he is usually the assiduous companion of his spouse, although he does not participate in her toils. They are late summer insects, and form their burrows upon banks having a southern aspect; these they excavate deeper than does *Andrena*, and smooth and polish them internally. They generally prefer spots intertangled with shrubs, and at the mouth of the cylinder they tunnel they heap up the extracted soil, to use a portion for closing it when their task is accomplished. In the course of this process, especially if a cloud pass over the sun, they will come forward to the aperture. They collect large quantities of pollen, for which the hair upon their posterior tibiæ and plantæ is excellently well adapted both by its length and the additional storing power it possesses in each individual hair being spirally twisted, although they are unprovided with the furniture of hair upon the femora and coxæ found in the genus *Andrenæ*. Thus nature likes to vary its mode of accomplishing the same object. The details of their nursery processes are not known. For their protection their sting is very virulent, and also actively employed, as they have many enemies, especially amongst the fossorial *Hymenoptera*, whom they stoutly resist to the extent of their strength. We are not aware of any special parasites that infest them. They are semigregarious in their habits, for where they occur any quantity of them may be taken. They are found in their season in the southern counties, the Isle of Wight, and in several parts of Kent and its eastern coast, and even as near London as Charlton. They seem to prefer the composite flowers, having a great liking for the bastard Hawkweed and the Dandelion. A fine series of them forms a great ornament to a collection.

Subfamily 2. APIDÆ (Normal Bees), *Latreille.*
Syn. APIS, *Kirby.*
Tongue *always folded back in repose.*
Maxillary palpi *varying in the number of the joints.*

Section 1. *Solitary.*

Subsection 1. SCOPULIPEDES (brush-legged).

a. *Femoriferæ* (collectors on entire leg).

† *With two submarginal cells to the wings.*

Genus 9. PANURGUS, *Panzer.*

(Plate VI. fig. 1 ♂ ♀.)

APIS * *a*, *Kirby.*

Gen. Char.: HEAD transversely subquadrate; *ocelli* in a triangle on the *vertex*, which, as well as the *face*, is convex, the latter between the antennæ carinated as far as the clypeus; *antennæ* short, subclavate, the second joint of the flagellum considerably the longest, the remainder equal; *clypeus* slightly convex; *labrum* transversely quadrate, convex; *mandibles* acutely unidentate; *cibarial apparatus* long; *tongue* half its entire length, gradually acute, and fringed laterally with delicate hair; *paraglossæ* slender, acute, membranous, not quite half the length of the tongue; *labial palpi* more than half the length of the tongue, the basal joint longer than the two following, the remainder gradually decreasing in length, all conterminous; *labium* half the length of the cibarial apparatus, broad; *maxillæ* slender, subhastate, as long as the tongue; *maxillary palpi* six-jointed, the basal joint robust, subclavate, as is the second joint, but more slender, the remainder filiform, gradually declining in length. THORAX oval; *prothorax* inconspicuous; *mesothorax* with a deep central groove; *bosses* protuberant; *scutellum* and *post-scutellum* lunulate; *metathorax* gibbous; *wings* with the marginal cell slightly appendiculated, two submarginal cells and a third commenced, the second receiving both the recurrent nervures, the first close to its commencement and the second beyond its centre; the *legs* densely pilose, the posterior pair having their *coxæ* and *trochanters* beneath, their *femora* in front, above, the *tibiæ* and *plantæ* all round, covered with long hair; *claws* bifid. ABDOMEN ovate, the base subtruncate, the basal segment having a deep central impression at its base, the fifth segment fringed with short dense hair, the terminal segment with a triangular plate carinated in the centre, and fimbriated laterally, and all very slightly constricted.

The MALE scarcely differs, except in having the *head* rather more globose and more pubescent; and the *legs*, although still hairy, much less so than in the female.

NATIVE SPECIES.

1. *Banksiana*, Kirby, ♂ ♀.   4-5¼ lines.
   ursinus, Curtis, iii. 101.   (Plate VI. fig. 1 ♂ ♀.)
2. *calcaratus*, Scopoli, ♂ ♀.   3-4 lines.
   ursinus, Kirby.

GENERAL OBSERVATIONS.

Πανοῦργος signifies *one excessively industrious*, at least as it is applied here, although it has other less meritorious meanings, but these insects can scarcely be considered more energetic than any of their associates; perhaps the contrast made between the bright yellow pollen and their lugubrious vestment might give the idea of very active collecting, they being usually, upon returning from their foray, almost entirely disguised in the produce of their excursion. They are rather remarkable insects from their intensely black colour and their compact active forms; their square head and short clavate antennæ give them a sturdy business-like appearance. They also are silent on the wing, but being at the very van of the present subfamily, forming as it were the advanced picket of the *Apidæ*, it may be considered suitable that they should retain, by way of partial disguise, some of the characteristics of the preceding subfamily. In many respects, therefore, they closely approach *Dasypoda*: thus their legs are similarly furnished with hair, relatively as long and having the same spiral twist, and their whole habit is that of one of the *Andrenidæ*, excepting that their clavate antennæ, and the folding of their tongue in repose, separate them from that subfamily. They are local insects, but extremely abundant when fallen upon. I used to find the first species upon an elevated plateau, on the south side overhanging the Vale of Health and its large pond at Hampstead. Every Dandelion, for a wide circuit in the vicinity, was crowded with individuals—assiduously collecting, in the case of females, but basking in sunny indolence, and revelling in the attractions of the flower, in the case of males, and, at the same time, their burrowing spot, which was not larger than half-a-dozen square yards, was swarming with them, coming and going, burrowing and provisioning. Very numerous, but not so numerous as themselves, were their pretty parasite, the *Nomada Fabriciana*, fine specimens of both sexes of which I have constantly captured; and a remarkable singularity pertaining to the latter is, that some seasons it would totally fail, and another season present itself sparsely, when, after these lapses, it would recur in all its primitive profusion, although the *Panurgus* was every season equally present. Both these insects are found during the months of June and July, especially about the middle of the former. In their burrows, which they perforate vertically, they usually enclose about six cells, each being duly provisioned and the egg deposited, when each is separately closed and the orifice of the cylinder filled up. This species is also found in Kent and Surrey, and I have no doubt they might be discovered in most of the southern counties. The smaller species, which is a good deal like a little *Tiphia*, is remarkable for the peculiarity of the male having a projecting process upon its posterior femora, whence it derives its specific name, *calcaratus*, which is hardly consistent, as it is not quite the right place for a spur. This smaller species is also found in Kent, Hampshire, and at Weybridge, in Surrey, and in the Isle of Wight. As well as in the *Leontodon*, it likes to repose in the flowers of the Mouse-ear Hawkweed (*Hieracium*).

b. *Cruriferæ* (collectors on the shanks and tarsi).

† *With two submarginal cells to the wings.*

Genus 10. EUCERA, *Scopoli*.

(Plate VI. fig. 2 ♂ ♀.)

Apis ** *d* 1, Kirby.

Gen. Char.: HEAD transverse; *vertex* concave; *ocelli* in a curve, and very high up; *face* flattish; *clypeus* very convex, hirsute, and fimbriated; *labrum* transverse-ovate, and emarginate in front; *mandibles* very obtusely and inconspicuously bidentate; *tongue* very long and slender, and gradually acuminating, transversely striated; *paraglossæ* slender, membranous, very acute, and about two-thirds the length of the tongue; *labial palpi* membranous, and about the length of the paraglossæ, the basal joint linear, broad, longer than the rest united, the second about half its length and acuminate, the two terminal ones are very short and equal, and articulate within the apex of the second joint; *labium* less than half the length of the tongue, its inosculation concave; *maxillæ* two-thirds the length of the tongue, subhastate; *maxillary palpi* six-jointed, short, less than one-third the length of the maxillæ, the basal joint robust, the rest filiform, and gradually decreasing in length and substance. THORAX very pubescent, which conceals its divisions; *metathorax* truncated; *wings* with two submarginal cells, the second receiving both the recurrent nervures, one near each of its extremities; *legs* setose, especially the tibiæ and plantæ, which, in the posterior pair is very dense on the exterior of the tibiæ, and both externally and internally upon the plantæ, the following joints of the posterior tarsi inserted beneath, and within the extremity of their plantæ; the claw-joint being longer than the two preceding, and the *claws* acutely bifid. ABDOMEN oval, convex above, subtruncate at the base, where it is thickly pubescent, the other segments glabrous on the disk; the fifth segment fimbriated with decumbent short hair, and the terminal segment having a central triangular plate at the sides of which it is rigidly setose.

The MALE differs in having the *antennæ* longer than the thorax, filiform, but with their several joints curved, the curvature increasing towards the terminal joints, the integument of the whole of the flagellum consisting of a congeries of minute hexagons, the edges of which are all raised, and the whole resembling shagreen; the legs have the usual sexual slighter and extended development, and are necessarily less setose; it is also deficient in the transverse whitish bands of decumbent hair upon the abdomen, which is more densely pubescent on the first and second segments; and the four terminal joints of the posterior tarsi are conterminous with their plantæ.

NATIVE SPECIES.

1. *longicornis*, Linnæus. 6–7 lines. (Plate VI. fig. 2 ♂ ♀.)
longicornis, Kirby.

GENERAL OBSERVATIONS.

This genus derives its name from the great length of the antennæ in the male,—εὖ, *good* or *great*, κέρας, *horn*. The name of the genus is usually given from some female characteristic, or from a peculiarity common to both sexes, or irrespective of any direct application, but here we find it deduced from a feature exclusively masculine. Instances of the first class we see in *Colletes*,

Halictus, Andrena, Dasypoda, Panurgus, Saropoda, Ceratina, Cœlioxys, Chelostoma, Heriades, Anthocopa, and *Apathus;* of the second class we have *Prosopis, Sphecodes, Macropis, Anthophora, Nomada, Melecta,* perhaps *Epeolus,* according to Latreille's idea, *Stelis, Anthidium, Osmia,* and *Bombus;* the third class comprises in our series merely *Cilissa,* and in this series the male characteristics that have suggested the name are just as few, being limited to the present genus. But the males among the bees exhibit in many cases strong and striking peculiarities which distinguish them from their partners. Exclusively of the general distinction expressed in their organic difference by the possession of one additional joint to the antennæ and one more segment to the abdomen than is exhibited in the females, we find in many cases in these two parts of their structure very marked singularities. Great sexual differences in the length of the antennæ are not restricted to the present genus; in fact, in most of the genera, this is the first striking feature, but which becomes conspicuously so in some species of *Sphecodes,* in most of the *Halicti,* in some *Nomadæ,* in *Chelostoma, Osmia, Apathus,* and *Bombus.* In *Eucera* and *Sphecodes,* each joint of the flagellum is slightly curved, and in the former the surface of those joints appears compounded of hexagons. In *Chelostoma* the antennæ, besides being longer than in the female, are also very much slighter and slightly compressed, and have a structure capable of curling upon itself; in the female of this genus the organ is clavate; and in *Osmia,* besides their length, in one species the male has a fringe of hair attached to one side along the whole of the organ. In other cases, where the antennæ are not remarkably longer in the male they have extra development by becoming thicker, as in *Melecta;* and in *Megachile* the terminal joint of their antennæ is laterally dilated and compressed. In scarcely any case are they geniculated at the scape in the male, as they are in the female. The other genera with clavate antennæ have the same structure in both sexes, as in *Panurgus* and *Ceratina.* Remarkable peculiarities in the terminal ventral segment or segments of the male may be found most conspicuously developed in *Halictus, Cœlioxys, Anthidium, Chelostoma, Heriades, Osmia, Apathus, Bombus,* and *Apis.* In *Cœlioxys* and *Anthidium,* and some of the *Osmiæ,* this sex is further furnished with a series of projecting spines, processes, or serrations at the apex of the terminal dorsal segment. In *Chelostoma,* the ventral structure of the male is very singular, the apex being adapted to a mucro at the base which permits the insect to curl up this portion of the body similarly to its antennæ, the furcated extremity of the abdomen fitting, when thus folded, upon the mucro. It is as well to draw observation to these peculiarities, which give additional interest to the study of the group.

The genus *Eucera* appears in May and June. In some parts they are found in large colonies; although I have seen them abundant I never found them in this gregarious condition, and I have usually discovered them frequenting loamy and sandy soils; they burrow a cell six or eight inches deep, form an oval chamber at its extremity, which as well as the sides of the cylinder leading to it they make extremely smooth, and by some process prevent its absorbing the mixture of honey and pollen which they store for the supply of the larva, and each contains but one young one. These, having full fed, lie in a dormant state throughout the winter and do not change into pupæ until mid-spring, and speedily transform into the imago, which, until fully matured, is closely in every part and limb covered with a thin silky pellicle, wherein it lies as in a shroud, but at its appointed time, regulated by some influence of which we have no cognizance, active life becomes developed, it then casts off its envelope and comes forth to revel in the sunshine, in close companionship with a partner which its instinct promptly teaches it to find. The largest of our native *Nomadæ* is its parasite the *N. sexcincta*, and which seems wholly restricted to it, but which is often even rare in places where the *Eucera* abounds. The female, like those of the rest of the bees, is no time-waster, but flies steadily to and fro in her occupation of provisioning her nest, and the male often accompanies her in these expeditions, gallantly winging about with extreme velocity as if to divert his sedulous companion in the fatigue of her toil, by his evolutions and his music, which is very sonorous. And on a fine May day it is extremely pleasant in a picturesque situation to sit and watch the operations of these very active insects. In their recent state, when just evolved from the nidus, they are very elegant, being covered with a close silky down, which labour and exposure soon abrades. It is said that this bee deserts her nest when she finds the stranger's egg deposited on the provender laid up in store, or when she meets with the *Nomada* within, which sometimes lays two eggs in one cell. To this she does not deliver battle, as does the *Anthophora* to *Melecta*, but patiently vacates the nest, leaving it to the service of the parasite, which is also supposed to close it herself, having been caught with clay encrusted upon her posterior legs. For the accuracy of this supposition I cannot vouch, never having observed the circumstance, nor have I seen reason to abandon the idea that the parasite has no instinct for labour of any kind,—the presence of the clay being, I expect, merely accidental, for it is notorious that these insects have an overruling predilection for keeping themselves extremely clean.

---

†† *With three submarginal cells to the wings.*

Genus 2. ANTHOPHORA, *Latreille.*

(Plate VI. fig. 3, and Plate VII. fig. 1.)

APIS ** *d*, 2 *a*, Kirby.

Gen. Char.: HEAD transverse, nearly as wide as the thorax; *vertex* depressed; *ocelli* placed in a curved line upon its posterior margin; *antennæ* short, subclavate, basal joint of flagellum globose, its second joint longer than the scape, very slender, the rest of the joints subequal; *face* flattish; *clypeus* protuberant; *labrum* quadrate, convex; *mandibles* distinctly bidentate and obtuse; *cibarial apparatus* very long; *tongue* very long, transversely striated, and with a small knob at the extremity; *paraglossæ* about one-third the length of the tongue, acuminate; *labial palpi* slender, more than half the length of the tongue, membranous, the basal joint as long again as the remainder, the second joint very slender and very acute; the two terminal joints very short and subclavate, inserted before the extremity of the second joint; *labium* short, one-fourth the length of the tongue, its inosculation concave; *maxillæ* hastate, not so long as the tongue; *maxillary palpi* one-third the length of the maxillæ, six-jointed, the basal joint very robust, the rest filiform, the second the longest,

and all the rest decreasing in length and substance.
Thorax oval, densely pubescent, which conceals its
divisions; *metathorax* truncated; *wings* with three
submarginal cells, closed, the second receives the first
recurrent nervure in its centre, and the third, which
bulges externally, receives the second at its extremity;
legs setose, the exterior of the posterior *tibiæ* and
plantæ moderately so, and the interior of the latter also
densely setose; the second joint of the posterior *tarsi*
inserted beneath and within the termination of their
plantæ; the claw joint longer than the two preceding;
claws bifid, the inner tooth distant from the external.
Abdomen ovate, subpubescent, the fifth segment densely
fimbriated and the terminal segment with an emarginate
appendage.

In the males the antennæ are very similar, but the
mandibles are more acutely bidentate, and with the
exception of the form of the legs, the general aspect is
like the female; the legs, although setose, are less conspicuously so, the intermediate tarsi in the first section
of the genus being longer than the rest of the entire leg,
and are fringed externally with very long hair, or it is
restricted to the plantæ of that leg and then it is short
and very rigid; the entire limb stretched out extends
beyond the widest expansion of the superior wings.
The abdomen is also less retuse than in the female, at
its basal segment.

In the second division of this genus, of which *Anthophora furcata* may be considered to be the type, the
general habit is precisely the same, but the insects are
not so pubescent, and there is a greater similarity between the sexes. The intermediate legs also, although
long in the male, are not so extremely long as they are
in the first section.

NATIVE SPECIES.

§ *Males with elongate tufted intermediate tarsi, and differing from female in colour.*

1. *retusa*, Linnæus, ♂ ♀. 6 lines. (Plate VI. fig. 3 ♂ ♀.)
   Haworthana, Kirby.
   Haworthana, Curtis, viii. 357.
2. *acervorum*, Fabricius, ♂ ♀. 6–8 lines.
   retusa, Kirby.

§§ *Males without elongate tufted intermediate tarsi, concolorous with their females.*

3. *furcata*, Panzer, ♂ ♀. 5–6 lines. (Plate VII. fig. 1 ♂ ♀.)
   furcata, Kirby.
4. *quadrimaculata*, Panzer, ♂ ♀. 4–5 lines.
   vulpina, Kirby.
   subglobosa, Kirby.

GENERAL OBSERVATIONS.

The name ἄνθος, φώρ φωρὸς, *flower-rifler*, would be as suitable for any other genus of bees, and therefore may be classed with those names which have no explicit signification.

The two divisions which our native species of this genus form, might very consistently constitute two genera, differing so much as they do both in habit and habits. In the first section the males totally differ from their females, the latter being black and the pubescence of their partners fulvous, and whose intermediate legs are so much longer, and are decorated besides with tufts of hair upon their plantæ, neither peculiarity being found in those of the second section, which conform more regularly to the ordinary type of structure. The first section also nidificate gregariously, forming enormous colonies which consist of many hundreds; whereas the second are solitary nidificators, and at most half-a-dozen may be found within as many square yards of territory, and one species, the *A. furcata,* diverges considerably from the ordinary habits of the genus, and closely approaches those of the foreign genus *Xylocopa,* but its structure necessarily retains it within the boundaries of the genus. All these insects exhibit the peculiar characteristic of the *Scopulipedes,* in the insertion of the second joint of the posterior tarsi at the very bottom of their plantæ, conjunctively with the polliniferous scopa, placed externally upon their tibiæ and plantæ, in which characteristics the Andrenoid *Macropis* remarkably resembles them, and which I have noticed in my remarks upon that genus.

The first section burrows in banks, where their colonies are extremely numerous. In the tunnels which they form they construct several elliptical cells which they line with a delicate membrane of a white colour, formed by a secretion or saliva derived from the digestion of either the pollen or the honey which they consume. Each cell when formed is stored as usual, and the egg deposited, and then it is closed. There is but little variation in these processes among all the solitary bees, excepting in the case of the artisan bees and the more elaborate processes of *Colletes,* in which, however, the casing is merely thicker, arising from several layers of the coating membrane. The perfect insects make their appearance during the spring and summer months, their successive maturity being the result of the previous summer and autumn deposit of eggs. They pass the winter and spring in the larva state, and undergo their transformations into pupa and imago with but slight interval, and only shortly before the appearance of the perfect insect. When first presenting themselves they are certainly very handsome insects, and if carefully killed preserve their beauty for many years in the cabinet. I have found the *retusa*, Linn., (Kirby's *Haworthana*,) in enormous profusion at Hampstead Heath, indeed, so numerous were they, that late in the afternoon, upon approaching the colony, they, in returning home, would strike as forcibly against me as is often done by *Melolontha vulgaris* or *Geotrupes stercorarius*. In equal abundance I have found the *A. acervorum* at Charlton, where I have experienced a similar battery. This is the insect which Gilbert White, in his letters from Selborne, describes as having found in numbers at Mount Caburn, near Lewes, a spot I have often visited in my schoolboy days. This section is subject to the parasitism of the genus *Melecta*, whose incursions are very repugnant to them, and which they exhibit in very fierce pugnacity, for if they catch the intruder in her invasion they will draw her forth and deliver battle with great fury. I have seen both the combatants rolling in the dust, the combat and escape made perhaps easier to the *Melecta* by the load the *Anthophora* was bearing home. Upon the larva also of this bee it is said that the larva of the Heteromerous genus *Meloë* is nurtured; this I have never been able to verify, but I believe the fact is fully confirmed. This beetle is closely allied to the *Cantharides*, or blister-beetles, and it itself exudes a very acrimonious yellow liquid when touched or irritated. Two of the *Chalcididæ* also infest their larvæ, which they destroy; one is the *Melittobia*, named thus from its preying upon bees; it, like the majority of its tribe, is exceedingly minute, and of a shining dark green metallic colour. It is peculiar from having its lateral eyes simple, and in possessing besides three ocelli. The other genus is *Monodontomeris*, an equally small insect, which, although living upon the larva of *Anthophora*, is equally preyed upon by that of the *Melittobia*. The universal scourge, *Forficula*, is a great devastator of these colonies, where, of course, it revels in its destructive propensities.

The insects of the second division I have never been able to track to their burrows, but have always caught them either on the wing or on flowers, especially upon those of the common Mallow, and I have found both species all round London. They are said also to frequent the Dead Nettle (*Lamium purpureum*). The *A. quadrimaculata* burrows in banks, and its processes are scarcely different from those of the preceding species, only its habits are solitary. In flight it is exceedingly rapid, and thus much resembles *Saropoda*. But the *A. furcata* bores into putrescent wood, in which it forms a longitudinal pipe subdivided into nine or ten oval divisions, separated from each other by agglutinated scrapings of the same material, very much masticated, the closing of each forming a sharp sort of cornice; each of these cells is about half an inch in length, and three-tenths of an inch in diameter, the separations between them being about a line thick. These pipes or cylinders run parallel to the sides of the wood thus bored, an angle being made both at its commencement and its termination, and thus the latter permits the ready escape of the developed imago nearest that extremity, which being the first deposited, that cell being the first constructed, it

R

necessarily becomes the first transmuted, and thus has not to wait for the egress of all above it.

All these insects are usually accompanied by their partners in their flight, and their amorous intercourse takes place upon the wing.

Genus 12. SAROPODA, *Latreille.*

(Plate VII. fig. 2 ♂ ♀.)

APIS ** *d*, 2, *a*, Kirby.

Gen. Char.: HEAD transverse, as wide as the thorax, very pubescent; *ocelli* placed in a triangle, the anterior one low towards the face; *vertex* slightly concave; *antennæ* short, filiform, basal joint of flagellum globose, the second joint subclavate and the longest, the rest short and equal; *face* flattish, short; *clypeus* forming an obtuse triangle, slightly convex; *labrum* quadrate, with the angles rounded; *mandibles* obtusely bidentate; *cibarial apparatus* long; *tongue* very long and slender, but gradually expanding towards half its length and then as gradually tapering to the extremity and terminating in a small knob, its sides throughout being fimbriated with short delicate down; *paraglossæ* one-third its length, membranous, very delicate, and tapering to a point; *labial palpi* slender, membranous, the joints conterminous, the basal joint more than half the length of the tongue, the remainder short, the second the longest of these three, and all tapering to the pointed apical one; *labium* scarcely one-third as long as the tongue, rather broad, bifid at its inosculation; *maxillæ* nearly as long as the tongue, gradually diminishing from its basal sinus to a point at its extremity; *maxillary palpi* four-jointed, about one-third the length of the maxillæ, the basal joint short, robust, the second tapering from its base to the third joint, which is rather shorter and subclavate, the terminal joint slender. THORAX very pubescent, rendering its divisions inconspicuous; *scutellum* and *post-scutellum* lunulate and convex; *metathorax* truncated; *wings* as in *Anthophora*, with three marginal cells closed, the second forming a truncated triangle, and receiving the first recurrent nervure near its centre, the third bulging outwardly and receiving the second recurrent nervure at its extremity; *legs* very setose, especially the posterior tibiæ externally, and their plantæ both externally and internally, but the setæ are longer on the exterior of the joint, the second joint of these tarsi inserted beneath, and before the termination of their plantæ, the terminal joint longer than the two preceding; *claws* bifid, the inner tooth distant from the apex. ABDOMEN subovate, very convex, truncated at its base, where it is densely pubescent, the fifth segment fimbriated with stiff setæ, and the terminal segment having a central triangular plate with rigid setæ at its sides.

The MALE scarcely differs, excepting in the characteristic sexual disparities of slightly longer antennæ, and considerably longer intermediate tarsi, whose apical joint is very clavate.

NATIVE SPECIES.

1. *bimaculata*, Panzer. ♂ ♀. 4–5 lines. (Plate VII. fig. 2 ♂ ♀.)
bimaculata, Kirby.
rotundata, Kirby.

GENERAL OBSERVATIONS.

The name of this genus is as applicable to the subsection as to the genus itself, σάρος, *brush*, ποῦς ποδὸς, *a foot*, in allusion to their polliniferous posterior legs.

We have but one species, but it is very characteristic; for, although retaining several of the features of the second division of *Anthophora* (in the colouring of the face it participates with the males of both divisions), yet has it still a marked physiognomy of its own; it retains the normal colouring of bees generally, but its strongest distinction from that division of *Anthophora* is the shortness of the antennæ in the female, as in the length of the intermediate legs of the male it would seem to form a link between the two divisions, could a distinct genus stand in such a position, and would almost import the necessity of elevating that division to generic rank, as hinted at in the observations under *Anthophora*. In the large development of its claws it seems to point to an economy somewhat differing from that second division, but nobody appears to have traced it to its nidus. I have often captured it at Battersea upon the Mallow, together with *A. quadrimaculata*, but the singular velocity of its flight might indicate a very distant domicile,—in a few minutes it could traverse miles. The electrical vivacity and rich opaline tint of its eyes has been often observed, but this, unfortunately, fades with death; yet so marked is it that it has called forth the distinct observation of a Panzer and a Kirby. Besides the Mallow it has been observed to frequent the Heaths, and were its habits better known would be found, I have no doubt, to visit many other flowers, for Curtis took it in the Isle of Wight sleeping in the great Knapweed, *Centaurea scabiosa*. I have never caught it laden.

I have hazarded the conjecture in a different part of this work that the music of the bees might be attuned to a musical scale by associating the different species in the due gradation of their varying tones. Here we have one of the most musical of the tribe,—not a monotonous dull sleepy hum, but a fine *contralto*, the very Patti amongst the bees. But it is rapidity of motion which in them intensifies the note they chant, and the velocity of the flight of this insect is something remarkable. They dart about with almost the rapidity of a flash of lightning, and this swiftness of approach and retreat modulates their accents.

Under the head "Macropis" I have pointed to some strong resemblances between this genus and that.

Genus 13. CERATINA, *Latreille.*

(Plate VII. fig. 3 ♂ ♀.)

APIS **d 2, *a*, Kirby.

Gen. Char.: HEAD transverse, convex, glabrous; *ocelli* placed in a triangle on the vertex, which is, as well as the face, convex; *antennæ* short, subclavate, each inserted in a separate deep cavity in the centre of the face, the first joint of the flagellum globose, the second the longest of all and slender at its base, but all gradually enlarging to the extremity; *clypeus* very gibbous; *labrum* quadrate, convex; *cibarial apparatus* long; *tongue* long and tapering, and with a minute knob at its extremity; *paraglossæ* obsolete; *labial palpi* three-fourths as long as the tongue, the two first joints membranous and diminishing in width, the second joint rather shorter than the basal one and acute at its extremity, and externally before its termination the two very short terminal ones are inserted; *labium* half the length of the tongue, with a lozenge-shaped inosculation; *maxillæ* as long as the tongue, broad at the base, whence it abruptly acuminates to the slender apex; *maxillary palpi* six-jointed, filiform, the three first joints subequal, the three terminal gradually decreasing in length. THORAX oval, glabrous; *prothorax* inconspicuous; *mesothorax* with a central basal groove, the *bosses* conspicuous and shining; *scutellum* and *post-scutellum* lunulate; *metathorax* subtruncate; *wings* with three submarginal cells and a fourth slightly commenced, the second in the form of a truncated triangle, the third considerably larger than the second, and each receiving a recurrent nervure just beyond the centre; *legs* plumose but not densely so, the hair very long within the posterior tibiæ, but denser and shorter on its exterior; the *posterior plantæ* also plumose, and all the joints of the posterior tarsi conterminous; *claws* bifid. ABDOMEN glabrous, subclavate, very convex above and flat beneath, subtruncate at the base, and the basal segments slightly constricted.

The MALE scarcely differs, excepting in the *clypeus* being less gibbous, the *legs* not plumose, and the sixth segment of the *abdomen* carinated in the centre towards its extremity, and impending over the seventh, which is transversely gibbous, then depressed, and with an obtuse process at its extremity.

NATIVE SPECIES.

1. *cærulea*, Villers, ♂ ♀. 2-3 lines. (Plate VII. fig. 3 ♂ ♀.)
cyanea, Kirby.
2. *albilabris*, Fabricius, ♂ ♀. 2½ lines.

GENERAL OBSERVATIONS.

This genus is named from the presence of a little

horn between its antennæ, κερατίνη, *a horn*. Some foreign entomologists, especially Latreille and Le Pelletier de St. Fargeau, have considered it to be parasitical, but that it is not so we have the authority of the Marquis Spinola, of Genoa, confirmed by the testimony of Mr. Thwaites, a very accurate observer, in the vicinity of Bristol, where the insect is not at all uncommon, although extremely rare in most other parts, and consequently usually a desideratum to cabinets, from its great beauty both of form and colour, notwithstanding that it is so very small in size. It has also been found in other localities, as at Birchwood, where the late Mr. Bambridge used to take it, and as near London as Charlton, at both which places I have no doubt it might frequently be found were it carefully looked for, but the practised entomological eye is often wanting to detect an insect unless it be conspicuously present. Its usual nidus is a bramble or briar stick, from which it excavates the pith, and this it has been frequently observed doing, and both sexes have been repeatedly bred from such sticks. We have no notice of any peculiarity in its mode of forming its cells, which may resemble that of such wood-boring genera as *Chelostoma* and *Heriades*, although its structure would intimate a closer affinity to the habits of the exotic genus *Xylocopa*; nor is there extant any account of the process or time occupied in the development of its young. Spinola's notion, from not seeing the sufficiency of the hair upon the posterior tibiæ for the purpose, assumed that the pollen was conveyed home on the forehead and between the antennæ, he having caught an insect with some pollen accidentally incrusted there in the insect's honey-seeking excursion. The hair upon these legs is very sparse, it is true, but then it is very long, and the quantity of pollen required for the nurture of the larva is evidently small, from its having been observed that the store upon which the egg is deposited is semiliquid, thus preponderating in the admixture of honey.

That it has not been caught laden with pollen upon its legs has no weight against the fact of its non-parasitism, for it is not always that the excursions of bees are made for the purpose of collecting pollen. Honey is as necessary to their economy—and in this case perhaps more so—as pollen, and the only way to determine the fact of its carrying pollen, corroboratively, would be when knowing that one of these bees has visited a bramble stick—its presumptive nidus,—to watch the stick very patiently for the insect's return from every journey until it came back laden; the presence of pollen upon its legs would surely be indicated by the difference of its colour from the ordinary dark hue of the little labourer.

We have already noticed bees with metallic hues among the *Halicti*, and there are slight indications of it in some of the *Andrenæ*, for instance, in the *A. cinerea* and the *A. nigro-ænea*, etc., but in none hitherto so absolutely is it exhibited as in this genus. The prevalent colour of the bees, that is to say, the ground colour of the integument, and not the fleeting one of the pubescence, is black or brown, but here we have a positive metallic tinge, which we shall again come across in many shades and hues in the genus *Osmia*.

A second species of the genus was brought from Devonshire by Dr. Leach, and is in the collection of the British Museum, but no other specimens of the same species have since been found.

The only flower which it has been noticed that they frequent is the Viper's Bugloss (*Echium vulgare*).

Subsection 2. NUDIPEDES (naked-legged Cuckoo Bees).
a. *With three submarginal cells to the wings.*

Genus 14. NOMADA, *Fabricius*.

(Plates VIII., IX., X.)

Gen. Char.: HEAD transverse; *ocelli* in a triangle on the vertex; *antennæ* filiform, scarcely geniculated, the scape short, the basal joint of the flagellum subglobose, the second joint clavate, the remainder subequal; *face* flat, or slightly concave, carinated longitudinally in the centre between the insertion of the antennæ; *clypeus* subtriangular, convex, deflected at the lateral angles; *labrum* subcircular, very gibbous and protuberant; *mandibles* acute or subbidentate; *tongue* long, acute; *paraglossæ* about one-fourth its length, acute; *labial palpi* two-thirds the length of the tongue, the two basal joints membranous, the basal one as long as the rest united, and tapering to its extremity, the second joint less than half the length of the first, and not wider at its base than the apex of the first joint, and tapering like that to its end, where it is acute, the third joint short, subclavate, and the terminal one half the length of the preceding, very slender and linear; *labium* about one-half the length of the tongue, and at its inosculation produced obtusely in the centre; *maxillæ* subhastate, about the length of the tongue; *maxillary palpi* six-jointed, the basal joint short, robust, subclavate, the second the longest, and with the rest tapering in substance and diminishing in length to the extremity, the terminal joint being very little shorter than the preceding. THORAX ovate; *prothorax* inconspicuous, or distinct and angulated laterally; *mesothorax* glabrous, deeply punctulated; its *bosses* conspicuous and prominent; *scutellum* divided into two very prominent tubercles; *post-scutellum* linear, convex; *metathorax* with a triangular space at its base, and declining to the insertion of the abdomen; *wings* with three submarginal cells, and a fourth very slightly commenced, the first as long as the two following, and each of which receives a recurrent nervure about its centre; *legs* subspinose externally on the tibiæ, and not polliniferous; *claws* of tarsi small and not bifid. ABDOMEN oval, glabrous, shining; terminal segment triangular, with its sides ridged.

The MALE scarcely differs, excepting in sometimes being more profusely adorned with colour, but this is not always the case, the female being often the most ornate. There are very slight differences in the antennæ in the sexes, which may be readily associated together.

NATIVE SPECIES.

§ *With filiform antennæ.*

1. *sex-fasciata*, Panzer, ♂ ♀. 5–6 lines. (Plate VIII. fig. 3 ♂ ♀.)
 Schæfferella, Kirby.
 connexa, Kirby.
2. *Goodemana*, Kirby, ♂ ♀. 4–5 lines. (Plate VIII. fig. 1 ♂ ♀.)
 ? *succincta*, Panzer.
3. *alternata*, Kirby, ♂ ♀. 4–5 lines.
 Marshamella, Kirby.
4. *Lathburiana*, Kirby, ♂ ♀. 4–5½ lines. (Plate VIII. fig. 2 ♂. ♀.)

5. *varia,* Panzer, ♂ ♀. 4–4½ lines.
 varia, Kirby.
 fucata, Kirby.
6. *ruficornis,* Linnæus, ♂ ♀.
 ruficornis, Kirby.
 leucophthalma, Kirby.
 flava, Kirby.
7. *lateralis,* Panzer, ♂ ♀. 4–4½ lines. (Plate X. fig. 3 ♂ ♀.)
8. *ochrostoma,* Kirby, ♂ ♀. 4–4½ lines.
 Hillana, Kirby.
9. *signata,* Jurine, ♂ ♀. 4–5 lines. (Plate IX. fig. 1 ♂ ♀.)
10. *borealis,* Zetterstedt, ♂ ♀. 3½–5 lines.
11. *lineola,* Panzer, ♂ ♀. 4–6 lines.
 cornigera, Kirby.
 subcornuta, Kirby.
 Capreæ, Kirby.
 sex-cincta, Kirby.
12. *xanthosticta,* Kirby, ♂ ♀. 2–2¾ lines.
13. *flavo-guttata,* Kirby, ♂ ♀. 2–3 lines. (Plate IX. fig. 3 ♂ ♀.)
14. *furva,* Panzer, ♂ ♀. 2–2½ lines.
 rufocincta, Kirby.
 Sheppardana, Kirby.
 Dalii, Curtis.
15. *Germanica,* Panzer, ♂ ♀ 4 lines.
 ferruginata, Kirby.
16. *Fabriciana,* Linnæus, ♂ ♀. 3½–5 lines. (Plate IX. fig. 2 ♂ ♀.)
 Fabriciella, Kirby.
 quadrinotata, Kirby.
17. *armata,* Schaeffer, ♂ ♀. 5–5½ lines.

Kirbii, Stephens.

§§ *With subclavate antennæ.*

18. *Jacobeæ,* Panzer, ♂ ♀. 4–4½ lines. (Plate X. fig. 1 ♂ ♀.)
Jacobeæ, Kirby.
flavopicta, Kirby.
19. *Solidaginis,* Panzer, ♂ ♀. 3½–4 lines. (Plate X. fig. 2 ♂ ♀.)
picta, Kirby.
rufopicta, Kirby.
20. *Roberjeotiana,* Panzer, ♂ ♀. 3 lines.

GENERAL OBSERVATIONS.

This genus was named by Fabricius from the *Nomades*, a pastoral Scythian tribe, in allusion to the assumed wandering habits of the insects, and it is the fact indeed that they are usually found leisurely hovering about hedgerows, or the banks enclosing fields, or about the metropolis or nidus of any bee upon which they are parasitical. They are the gayest of all our bees, their colours being red or yellow intermixed with black, in bands or spots; they are also very elegant in form, which is after the type of that of the most normal *Andrenidæ*, and to which they have a further affinity by the silence of their flight, and by their parasitism upon many of the species of that subfamily. From their very general resemblance to wasps in colour they are often mistaken for wasps, and are popularly called wasp-bees, although they have none of the virulence of that vindictive tribe, for although all the females are armed with stings, they are not prompt in their use, or if roused to defence the puncture is but slight. In addition to their prettiness of colour and elegance of form, they have a further attraction in the agreeable odours they emit, sometimes of a balmy or balsamical, and sometimes of a mixed character, and often as sweet as the *pot-pourri*, and occasionally pleasantly pungent. A fine string of specimens of the several species is a great ornament to a collection, but to secure this in its perfection some care is required in the mode of killing them. Their colours are best permanently retained by suffocating them with sulphur, which fixes the reds and yellows in all their natural and living purity. My method was in my collecting excursions to convey with me a large store of pill-boxes of various sizes, and as I captured insects in my green gauze bag-net, I transferred them separately to these boxes. When home again I lifted the lids slightly on one side and placed as many as would readily go beneath a tumbler, and then fumigated them with the sulphur. This is a better plan than killing them with crushed laurel-leaves, for it leaves the limbs much longer flexible for the purposes of setting, whereas the laurel has a tendency to make them rigid, and this rigidity is extremely difficult to relax, whereas the setting of those killed with sulphur, if they are kept in a cool place, may be deferred for a few days, until leisure intervene to permit it, and even then if they become stiffened they are readily relaxed for the purpose.

A division might very consistently be established in the genus by the separation of those which have subclavate antennæ, and the segments of whose abdomen are slightly constricted; these also are more essentially midsummer insects, and usually frequent the Ragwort. This is the only genus of parasites amongst the true bees whose parasitism is directed exclusively upwards in the scientific arrangement; the parasitism of all the rest of the genera of *Nudipedes* bears upon the genera below them in the series. Some of the species of the *Nomadæ* attack more than one species or one genus, but the majority are strictly limited to but one genus and one species. The genera obnoxious to this annoyance are *Andrena, Halictus, Panurgus,* and *Eucera;* the latter two have but one of these enemies each, the *Nomada Fabriciana* infesting the *Panurgus Banksianus,* and the *N. sexfasciata* frequenting the *Eucera longicornis.* Under *Panurgus* I have alluded to the relative abundance of the parasite at the metropolis of its sitos. As far as known, the other species are thus distributed. Those frequenting several indifferently are the *Nomada alternata, Lathburiana, succincta,* and *ruficornis,* which are found to infest *Andrena Trimmerana, tibialis, Afzeliella,* and *fulva,* without displaying any choice; whereas others confine themselves to one sitos exclusively: thus *Nomada ochrostoma* limits itself to *Andrena labialis; N. Germanica* to *A. fulvescens; N. lateralis* to *A. longipes; N. baccata* to *A. argentata; N. borealis* to *A. Clarkella; N. Fabriciana* to *Panurgus Banksianus;* and *N. sexfasciata* to *Eucera longicornis.* Observation has not yet fully determined whither each species of *Nomada* conveys its parasitism; several infest the *Halicti,* especially the smaller species; the association of these it is difficult to determine; I have usually found several of the small *Halicti* burrowing together in the vertical surface of an enclosure bank, and several of the small *Nomada* hovering cautiously opposite, now alighting and entering a burrow, then retreating backwards and winging off. I lost patience in endeavouring to combine the species by the aid of blades of grass or slight straws thrust into the aperture, but the crumbling nature of the soil frustrated my wishes, and I abandoned the attempt. This field of observation is widely open to the exertions of observing naturalists, and the novelty of their discoveries would well reward the toil of the undertaking, for it would not be long before they gathered fruit.

Genus 15. MELECTA, *Latreille*.

(Plate XI. fig. 1 ♂ ♀.)

APIS ** *a*, Kirby.

Gen. Char.: HEAD transverse, scarcely so wide as the thorax; *ocelli* in a triangle on the vertex; *antennæ* filiform, rather robust, and but slightly geniculated, the scape not longer than the two following joints, the second joint of the flagellum the longest and clavate, the rest short, nearly equal, and the terminal one laterally compressed at its extremity; *face* flat, very pubescent; *clypeus* short transverse, lunulate, convex; *labrum* irregularly gibbous, obovate; *mandibles* strongly bidentate; *tongue* long, slightly expanding towards the middle and thence tapering to the extremity, and with a central line; *paraglossæ* scarcely half the length of the tongue, almost setiform, but robust at the base; *labial palpi* more than half the length of the tongue, the two first joints membranous and very slender, the first longer than the rest united, the second about half the length of the first, and terminating acutely, the third not more than one-fourth the length of the second, and inserted laterally before its termination, the fourth about as long as the third, and, like it, subclavate, both being more robust than the second; *labium* not half the length of the tongue, and acutely triangular at its inosculation; *maxillæ* subhastate, not quite so long as the tongue; *maxillary palpi* five-jointed, about one-third the length of the maxillæ, the basal joint clavate, short, and robust; the second elongate, subclavate, the remainder gradually but slightly diminishing in substance and length, the terminal not so long as the basal joint. THORAX very retuse, and its divisions scarcely distinguishable; *scutellum* bidentate; *metathorax* abruptly truncated; *wings* with three closed submarginal cells, the second the smallest, irregularly triangular, and receiving the first recurrent nervure just beyond its centre, the third submarginal considerably larger than the second, sublunulate, but angulated externally and receiving the second recurrent nervure about its centre; the *legs* robust and spinulose, especially the tibiæ externally (where they are very convex) and the femora beneath; the *claws* short, strong and bifid. The ABDOMEN conical, truncated, and retuse at its base, the apical segment with a central triangular plate ridged laterally, and fimbriated at its sides with strong setæ.

The MALE scarcely differs in personal appearance, excepting that its antennæ are more robust and its ornamental pubescence is more profuse, its posterior tibiæ very robust and almost triangular, and the terminal segment of its abdomen slightly emarginate and concave at its extremity.

NATIVE SPECIES.

1. *punctata*, Fabricius, ♂ ♀. 6 lines. (Plate XI. fig. 1 ♂ ♀.)

? *Atropos,* Newman.
? *Lachesis,* Newman.
2. *armata,* Panzer, ♂ ♀. 6-7 lines.
punctata, Kirby.
? *Tisiphone,* Newman.
? *Alecto,* Newman.
? *Clotho,* Newman.
? *Megæra,* Newman.

GENERAL OBSERVATIONS.

Named from μέλι, *honey,* λέγω, *I collect;* which is scarcely the case, for the parasites, although they may indulge in the luxury of honey as epicures, or resort to it as a repast, cannot be said to collect it, for it is only the labouring bees that truly collect it for the purpose of storing.

These insects are extremely handsome, their ground-colour being intensely black, brightly shining on the abdomen, upon the segments of which it is laterally ornamented with silvery pubescent tufts and spots; the black legs are also variously ringed with similar silver down. The great variation these spots and markings undergo—from what cause we know not—has induced several entomologists to consider them as distinct species. But the strongest varieties so rarely recur with identical ornaments, and as almost all can be closely connected together in a regular series by interlacing differences impossible to divide, it would be certainly incorrect, without stronger characteristics, to raise such fugitive variations to specific rank. Whether the curious spines of the scutellum which they possess furnish a more certain character is doubtful, for we find all such processes equally liable to variation in size and form. What can be the uses of these spines? They can hardly be for defence, although an entomologist has said that a male which he held endeavoured to pinch by that means. We find similar processes in the same situation in *Cœliox ys*, equally a parasitical genus; but the former genus infests the *Scopulipedes* and the latter the *Dasygasters*, whose economics are so very different, and thus it can hardly be supposed to have reference to habits. In *Epeolus* and *Stelis* the same part is mucronated, a tendency to which we see in the *Nomadæ* with subclavate antennæ. Under *Anthophora* I have given an account of the pugnacious spirit of these insects in their contests with the sitos, and it is necessary to be cautious in handling them, as they sting very severely. Our two native species are parasitical upon the two species of the first division of *Anthophora*, —those which are gregarious. The circumstance of *Melecta* being often caught with many of the extremely young larvæ of *Meloë* upon it seems to confirm the fact of this coleopterous insect preying upon *Anthophora*, as it may be thus assumed to prey simultaneously upon the larva of *Melecta*. I have never captured these insects upon flowers, nor can I trace what flowers they frequent, although Latreille tells us, in the name he has imposed, that they are honey collectors; but Curtis reports that he has found the genus upon the common furze or whin (*Ulex Europæus*).

Genus 16. EPEOLUS, *Latreille*.
(Plate XI. fig. 2 ♂ ♀.)
APIS ** *b*, Kirby.

Gen. Char.: BODY glabrous. HEAD transverse, vertex convex; *ocelli* placed in a triangle on its summit; *antennæ* short, linear, the joints of the flagellum subequal; *face* flat, carinated longitudinally in its centre between the insertion of the antennæ; *clypeus* transverse, lunulate, convex, margined anteriorly; *labrum* transversely ovate, with a small process in the centre in front; *mandibles* bidentate, the internal tooth minute, the external robust and broad; *tongue* rather long, more than twice the length of the labium, tapering to its extremity; *paraglossæ* short, about one-fourth the length of the tongue, broad at the base, and acuminate towards the apex; *labial palpi* more than half the length of the tongue, the basal joint longer than the three following, membranous, and gradually decreasing to the second, which is one-third the length of the first, and acute at its apex, where the third subclavate joint is articulated, the terminal joint considerably shorter than the third; *labium* not more than one-third the length of the tongue, and trifid at its inosculation, the central division being hastate; *maxillæ* subhastate, more than one-half the length of the tongue; *maxillary palpi* consisting of one robust short conical joint inserted in a deep circular receptacle. Thorax subglobose; *prothorax* conspicuous, with its lateral angles slightly prominent; *mesothorax* with its bosses prominent; *wing* scales large; *scutellum* transverse, gibbous, margined posteriorly, slightly mucronated laterally, slightly depressed in the centre, and impending over the *post-scutellum*, which is inapparent; *metathorax* abruptly truncated; *wings* with three submarginal cells, and a fourth feebly commenced, the first as long as the two following, the second subtriangular, and receiving the first recurrent nervure about its centre, and the third lunulate, and receiving the second recurrent nervure also about the centre; *legs* short, stout, the *tibiæ* slightly spinulose externally; *claws* very small, short, robust and simple. ABDOMEN obtusely conical, truncated at the base, its terminal segment triangular, and the lateral margins slightly reflected.

The MALE scarcely differs, excepting in the usual male characteristics, and that the apical segment of the abdomen is rounded and margined.

NATIVE SPECIES.

1. *variegatus*, Linnæus, ♂ ♀. 3–4 lines. (Plate XI. fig. 2 ♂ ♀.)
variegatus, Kirby.

GENERAL OBSERVATIONS.

It is difficult to assign a reason for the name of this genus, or to trace an applicable derivation from ἐπίαλος, for the insect in no way suits, either directly or by antiphrase, any of the significations of this word. It is one of the prettiest of our little bees, and is parasitical upon the *Colletes Daviesiana*, and it may be found in abundance wherever the metropolis of this species occurs. There is one special locality near Bexley, in Kent, a vertical sandbank within a few hundred yards of the village, where I have always found it in the spring months, and have there taken it as numerously as I wished. I have already alluded, in another part of this work, to the uniformly greater beauty of the parasitical bees, to those which they infest, and their exceedingly different appearance in every case excepting in that of the genus *Apathus*. We might have expected that they would have been disguised like these, the better to carry on their nefarious practices, but what can well be more dissimilar than *Epeolus* and *Colletes*, or than *Nomada* and all its supporters, and the same of *Melecta*, *Cœlioxys*, and *Stelis*. These facts puzzle investigation for a reason; nor will the perplexity be speedily solved. All that we can surmise is that there must be a motive for it, for wherever we successfully elicit her secret from the veiled goddess, we invariably find the reason founded in profound wisdom. In some cases the mystery seems devised to test our sagacity, but it cannot be so here, for the most palpable and plausible cause that would suggest itself in the supposition of its being for the guardianship and apprisal of the sitos is often contravened, as in this instance, by it and its parasite living in great harmony together, again by the desertion of its nidus by *Eucera* in favour of the parasite, although itself is a very much more powerful insect; but in the cases of *Panurgus*, *Halictus*, and *Andrena*, they all live well reconciled to the intrusion of the stranger's young, and this, without their enumeration, may be adopted as nearly the universal case. The hostility of *Anthophora*, previously noticed, is an almost insulated case of the contrary. The form of these insects does not promise much activity, and we accordingly find that they are slow, heavy, and indolent; yet they must be cautiously handled, for they sting acutely; but indeed it is not well ever to handle insects whose markings, as we find them in these, consist of a close nap, as evanescent as the down upon a plum, and of course the fingers carry it readily off, and disfigure the beauty of the little specimen. When their special habitat is not known they may often be found upon the blossoming Ragwort in the vicinity, or upon the Mouse-ear Hawkweed (*Hieracium murorum*) within whose flowers they are frequently observed enjoying their siesta.

b. *With two submarginal cells.*

Genus 17. STELIS, *Panzer.*

(Plate XI. fig. 3 ♂ ♀.)

APIS ** *c*, 1 β, Kirby.

Gen. Char.: BODY glabrous, much punctured. HEAD transverse, curving posteriorly to the thorax, where it is angulated laterally; *ocelli* in a triangle at the summit of the vertex; *antennæ* short, slender, filiform, scarcely geniculated, the scape about as long as the three first joints of the flagellum, all the joints of which are subequal but slightly increasing in length towards the apical one, which is a little compressed laterally; *face* entirely flat; *clypeus* transverse, rather convex; *labrum* elongate, convex; *mandibles* robust, tridentate, the external tooth considerably the stoutest; *cibarial apparatus* long, tongue three times as long as the labium, slightly inflated in the centre, and terminating in a small knob; *paraglossæ* very short, not more than one-sixth the length of the tongue and acuminate; *labial palpi* about two-thirds the length of the tongue, the two first joints membranous, the basal one the most robust, and both tapering to an acute apex, shortly before which the two very short subclavate terminal joints articulate; *labium* about one-third the length of the tongue, its inosculation trifid, the central division considerably the longest and truncated at its extremity; *maxillæ* subhastate, nearly as long as the tongue, acutely acuminated towards their apex; *maxiliary palpi* very short, two-jointed, the basal joint subclavate and slightly the longest, and inserted in a circular cavity, the terminal joint short ovate. THORAX subglobose; *prothorax* inconspicuous; *mesothorax* very convex; *scutellum* lunulate, very gibbous, and impending over the post-scutellum and metathorax, mucronated laterally; *metathorax* abruptly truncated; *wings* with two submarginal cells, and a third very slightly commenced, the two subequal, the second being the largest and receiving the first submarginal cell near its commencement and the second at the inosculation of the terminal transverso-cubital nervure; *legs* short, moderately stout, the *tibiæ* very slightly setose externally; *claws* short, bifid, the internal tooth near the external. ABDOMEN oblong, truncated at its base, very convex above and flat beneath, deflexed towards its extremity, and the terminal segment almost rounded, being very slightly produced in the centre and margined.

The MALE scarcely differs, excepting in the usual male characteristics, and by the apical segment being obsoletely tridentate.

NATIVE SPECIES.

1. *aterrima*, Panzer, ♂ ♀. 4–4½ lines.
   punctulatissima, Kirby,
2. *phæoptera*, Kirby, ♂ ♀. 4–4½ lines. (Plate XI. fig. 3 ♂ ♀.)
3. *octomaculata*, Smith, ♂ ♀. 3 lines.

GENERAL OBSERVATIONS.

The name of this genus may be derived from στελὶς, *a sort of parasitical plant*, perhaps *mistletoe*, if we could be sure that Panzer imposed it after being aware of the parasitical nature of these bees. It is true his book (the 'Revision') was published in 1805, and Kirby, who first intimated a suspicion of such cuckoo-like habits in some of the bees, published his in 1802; therefore it might have been given in allusion to that peculiarity of their economy, but it may also be from στηλίς, *a little column*, in application to their cylindrical form. In but few of the parasitical bees do we know the precise nature of their transformations, I have therefore been obliged to be silent upon this point of their natural history, and I have nothing to state of its nature in these, although I expect there is much uniformity with but slight modifications in all. The species of this genus are parasitical upon the *Osmiæ*; thus the *S. phæoptera* is found to infest the *O. fulviventris*, and the *S. octomaculata* intrudes itself into the nests of *O. leucomelana*, both of which occur tolerably abundantly near Bristol. I have no doubt that the south-west and west of England, if well searched, would yield many choice insects.

It is singular that bee parasitism does not prevail throughout all the genera of bees, some being subject to it and others not. Thus the genera *Colletes, Andrena, Halictus, Panurgus, Eucera, Anthophora, Saropoda, Megachile, Osmia,* and *Bombus* have all parasites, whereas the genera *Cilissa, Macropis, Dasypoda, Ceratina, Anthidium, Chelostoma, Heriades, Anthocopa,* and *Apis* have none, as far as we yet know; and some of the genera of parasites frequent two or more genera indifferently, whilst others are restricted to a single one; also some of the species of the parasitical genera infest indifferently several of the species of the genus to which their parasitism is mainly limited; other species have a more circumscribed range and do not visit the nests of more than a single species. What law may control all these seeming anomalies we cannot discover,—it may possibly be scent that guides them, and this may control their parasitism by indicating the species they are taught by their instinct to be most suitable from the quality of the pollen with which it supplies its own nest, to be that which is best adapted for the nurture of their young. It is not likely that we shall very speedily lift the veil from these mysteries, but they are suggestive of observation which in seeking one thing may fall upon another equally interesting.

I have usually caught these insects settled upon the leaves of shrubs, especially of fruit bushes, particularly that of the black currant, upon which, in a favourable locality, many bees, as well as numerous small fossorial *Hymenoptera* may be found in genial weather. I have never caught them upon flowers, nor do I know what flowers they frequent. The end of May, if warm, and throughout June, they are usually found most abundantly.

Genus 18. CŒLIOXYS, *Latreille*.

(Plate XII. fig. 1 ♂ ♀.)

APIS ** *c* 1 *a*, *Kirby*.

Gen. Char.: BODY subglabrous. HEAD transverse, concave posteriorly to fit the anterior portion of the thorax; *ocelli* in a triangle on the vertex; *antennæ* filiform, short, subgeniculated, the basal joint of the flagellum globose, the second subclavate, and all from the second subequal, the terminal joint compressed laterally; *face* flat, very pubescent; *clypeus* ovate, concavely truncated in front, its surface convex; *labrum* oblong, with its sides parallel, but with lateral processes at its articulation; *mandibles* broad, quadridentate; *cibarial apparatus* long, the *tongue* very long, nearly three times the length of the labium, linear but slightly inflated in the centre, and thence tapering to its extremity, and slightly covered with a very short down; *paraglossæ* wholly wanting; *labial palpi* membranous, the two first joints long, the second slightly the longest, and both tapering to the extremity of the second, which is acute, and has the third joint, which is very short and subclavate, articulated before the extremity, with the terminal one of equal length, and rounded at the apex, appended to it; *labium* about one-third the length of the tongue, its inosculation trifid and equal, and the central division acute; *maxillæ* subhastate and acuminate, not quite so long as the tongue; *maxillary palpi* very short, three-jointed, the basal joint the smallest, the second the most robust, and the terminal one ovate. THORAX subglobose; *prothorax* inconspicuous; *mesothorax* convex; *wing-scales* large; *scutellum* produced horizontally, and impending over the post-scutellum and metathorax, and having at each lateral extremity an acute, slightly-curved tooth projecting backwards; *metathorax* abruptly truncated; *wings* with two submarginal cells and a third commenced, the first slightly the longest, the second receiving both the recurrent nervures, the first near its commencement, and the second close to its termination; *legs* slender, spinulose externally on the tibiæ; *claws* rather long, slender, and simple. ABDOMEN very conical, truncated at the base, its segments slightly constricted, the apical one long, superficially carinated longitudinally in the centre, and much deflexed.

The MALE scarcely differs, excepting that the whole of the front of the head is more densely pubescent; the mandibles are deeply, acutely, and nearly equally tridentate, the terminal segment of the abdomen is variously mucronated or toothed at its apex, these processes pointing backwards, and the penultimate segment is more or less produced laterally.

NATIVE SPECIES.

1. *conica*, Linnæus, ♂ ♀. 4–5 lines.
   quadridentata, Linnæus, ♂.
   quadridentata, Kirby, ♂.
2. *simplex*, Nyland, ♂ ♀. 5 lines.
   conica, Kirby.
   conica, Curtis, viii. 349.
   Sponsa, Smith, ♂.
3. *umbrina*, Smith, ♂ ♀.
4. *rufescens*, St. Fargeau, ♂. 4–6 lines.
5. *vectis*, Curtis, ♂ ♀. 5–6 lines. (Plate XII. fig. 1 ♂ ♀.)
6. *inermis*, Kirby.

GENERAL OBSERVATIONS.

This genus is named from κοιλία, *belly*, ὀξύς, *acute*, in application to the conical abdomen of the female. The insects of this genus are parasitical upon the genera *Megachile* and *Saropoda*. Thus, *C. simplex* infests *M. circumcincta*; *C. rufescens*, *M. Willughbiella*; *C. vectis*, *M. maritima*; and *C. umbrina* is parasitical on *Saropoda bimaculata*. Linnæus, from the different appearance of the two sexes made two species of them, and from the circumstance of his having described first the male as *Apis quadridentata*, this, by the law of priority, supersedes the name of *C. conica* as the name of the species, which is its female, and which he next described, and thus that sex, whose form Latreille adopted as typical of the genus, is in the series of species totally superseded and reduced to a synonym. The species of this genus are extremely difficult to separate from each other, no tangible character presenting itself conspicuously, although the Swedish entomologist Nylander supposes he has found one in the plates of the apical segment of the abdomen, especially those of the venter, in which he detects both a difference of form and a difference of relative length to that of the superior plates, and in the males he assumes that the teeth of the apical segment are constant characters. Not having had sufficient opportunity since this supposed discovery was made, for the examination of a great multitude of specimens, for it is only upon such an investigation that it can be firmly based, I cannot speak corroboratively upon the point, but it is very possibly a correct solution of the difficulty.

The peculiarity of these spines at the apical segments of the abdomen of the males is remarkable, they being straight projecting processes, or they have even a slight upward bearing. In the males of *Anthidium* and *Osmia* we observe spines also arming the apex of the last segment, but in these we can trace an evident use, both from the downward curvature of the abdomen itself, and that same tendency also in the spines. But in the insects of this genus they have not the same conspicuously apparent object, the abdomen itself even having an upward curvature, or rather a greater facility for turning upwards than downwards. These insects appear to be most abundant in the midland and southern counties, and, according to Curtis, they are numerously found at the back of the Isle of Wight. I have usually taken them on the wing and never on a flower, and I do not know the plants which they may prefer.

Subsection 3. DASYGASTERS (convey pollen on the belly).

All with two submarginal cells to the wings.

Genus 19. MEGACHILE, *Latreille.* (Leaf-cutters.)

APIS ** *c* 2 *a*, Kirby.

(Plate XII. fig. 2 and 3 ♂ ♀.)

Gen. Char.: HEAD as wide as the thorax, flat and broad on the *vertex*, where, on the anterior edge, the *ocelli* are disposed in a triangle; *antennæ* shortish, filiform, geniculated; *scape* about as long as two first joints of flagellum, which increases both in length of joints and their substance from base to apex, the terminal one being the longest, and longitudinally compressed; *face* and *clypeus* very pubescent, concealing their divisions; *clypeus* transversely lunulate, scarcely convex; *labrum* longitudinally slightly convex and oblong, with the sides parallel; *mandibles* broad, widening outwardly, irregularly quadridental, the two inner teeth obtuse; *cibarial apparatus* moderately long; *tongue* more than twice the length of the labium, tapering from the base to the apex, where it terminates in a minute knob; *paraglossæ* very short, scarcely one-sixth the length of the tongue, coadunate at the base and acuminate at the apex, where, in repose, they lap round the base of the tongue; *labial palpi* three-fourths the length of the tongue, the two basal joints long, subequal, membranous, linear, slightly tapering to the acute apex of the second, where the third subclavate joint articulates just before its termination, and conterminous with which is the fourth, shorter than the third, but also subclavate; *labium* not quite half the length of the tongue, with a long subobtuse process in the centre of its inosculation; *maxillæ* subhastate, and very acuminate, nearly as long as the tongue; *maxillary palpi* very short, two-jointed, the basal joint the shortest, and the terminal one obtuse at its apex, where it is furnished with brief setæ. THORAX subglobose, pubescent, the pubescence almost concealing its divisions; *prothorax* inconspicuous; *mesothorax* convex, subglabrous on the disk; *scutellum* lunulate, convex; *metathorax* truncated; wings with two submarginal cells, the commencement of a third slightly indicated, the two complete ones nearly equal, the second of which receives both the recurrent nervures, one towards each extremity; *legs* robust, very setose; the posterior *tibiæ* slightly curved longitudinally, concavo-convex, broad at the extremity; all the *plantæ* as long as their tibiæ and as broad at the base but decreasing at the apex to the width of the following tarsal joints, the anterior pair fimbriated externally, and the posterior pair clothed, on the inner surface, with a dense, short brush, the three following joints short, subequal, the claw-joint as long as the three, and the *claws* with a broad basal inner tooth. ABDOMEN ovate, with parallel sides, convex above, truncated and concave at its base to fit the metathorax, distended horizontally in length, or with an upward curve, the four first segments slightly constricted, and their edges usually clothed with decumbent down; the terminal segment obtusely pointed and slightly depressed transversely towards its extremity; the *ventral segments* commencing with the second, clothed with parallel layers of moderately long, straight setæ, which in each parallel arc of equal length, but those on the fifth segment are the shortest, upon all of which the insect conveys the pollen it collects.

The FEMALES of the second division of the genus scarcely differ.

The MALES of the first division differ in having the *head* slightly larger and squarer above; the *antennæ* very slightly longer; the *mandibles* more acutely tridentate, with a distinct powerful basal tooth beneath, terminating the concavity of the organ; the anterior *femora, tibiæ,* and joints of their *tarsi,* excepting the terminal one, concavo-convex, the four first joints of the latter distended laterally, and edged with a dense fringe of setæ, the distension of these joints is widest at their articulation with the tibiæ and they decline in length to the claw joint which is long; the *claws* bifid; the interior claw acute, but remote from the apical one; the posterior *femora* are very robust, their *tibiæ* much curved, robust, almost triangular, and externally very convex; their *plantæ* almost glabrous, not so long as the three following joints, externally rather twisted, and beneath furnished with a dense brush of long stiff hair.

In the second division of the genus the males are destitute of the distension of the anterior tarsi, these being instead densely fimbriated externally; the legs in them are much less robust, and more closely resemble those of their females.

NATIVE SPECIES.

§ *Anterior tarsi of males much dilated.*

1. *Willughbiella,* Kirby, ♂ ♀. 5–7 lines.
2. *maritima,* Kirby, ♂ ♀. 6–7 lines. (Plate XII. fig. 2 ♂ ♀.)
3. *circumcincta,* Kirby, ♂ ♀. 4½–5½ lines.

§§ *Anterior tarsi of males not dilated.*

1. *ligniseca,* Kirby, ♂ ♀. 5–7 lines.
2. *centuncularis,* Linnæus, ♂ ♀. 4–6 lines.
   centuncularis, Kirby.

3. *argentata*, Fabricius, ♂ ♀. 3–4½ lines. (Plate
Leachella, Kirby. [XII. fig. 3 ♂ ♀.)
Leachella, Curtis.
4. *odontura*, Smith, ♂. 4½ lines.

GENERAL OBSERVATIONS.

Named from the great development of the labrum, μέγα *large*, χεῖλος *lip*, which is characteristic of all the *Dasygasters*, and also of some of the proximate *Nudipedes*, those parasitical upon them, *Stelis* and *Cœlioxys*, and which, too, resemble the sitos in the expansion and dentated formation of their mandibles, although they do not use them for the same purposes; this again exhibits an analogy of structure, that appears in the parasite to be merely corroborative of identity of existence.

These are more essentially summer insects than the majority of the preceding genera, although some of them present themselves with genial spring weather. The genus may be separated into two distinct divisions by the peculiar dilatation of the tarsi of the males of some of the species, but such division is not indicative of a difference of habits, as is distinctly the case in the genus *Anthophora*, and in which these combined circumstances Mr. Kirby suggested as acceptable for generic division, or, as he called it, the institution of another *family*. But in these we find in both divisions both wood-borers and earth-tunnelers, and some species are indifferently either as suits their accidental convenience. The general appearance of the insects is more that of ordinary bees, and the sexes are more approximate in their habit than is usually the case.

With this genus commences essentially those designated as artisan bees, although *Colletes* might very suitably come under that denomination. The species themselves of the genus are called leaf-cutters, from the habit they have of cutting pieces from the leaves of various shrubs and trees, for the purpose of lining their nests. The description of the operations of one species will apply precisely to that carried on by all, the occasional difference between them being the selection of the leaves of distinct plants; and it will exhibit the patient industry and perseverance with which these little upholsterers carry on their labours.

Thus *M. centuncularis*, the type of the genus, burrows in decaying wood or in brick walls, and sometimes also in the ground, and makes use of the cuttings of rose leaves, —not the petals,—and the leaves of the annual and perennial Mercury (*Mercurialis annua* and *M. perennis*). The *M. ligniseca* bores into sound Oak and the Mountain Ash, as well as into putrescent Elm, and uses Elm leaves to line its nests, sometimes called *centunculi* from their being as it were patched together. This is the largest of all our species, and is found very abundantly everywhere around London frequenting the flowers of the Thistle. The *M. argentata*, Fab., or *Leachella* of Kirby, is perhaps the prettiest of all the species, and forms its tunnels in sandbanks. I do not know what leaves this species selects, which used to be extremely rare, indeed for a long time only known by the specimen in the British Museum, until that ardent entomologist the Rev. F. W. Hope; to whom the University of Oxford owes its superb entomological collection, brought it in abundance from Southend, where, during his brief annual stay at his residence there, he used to find it in the grove which runs under the cliff edging the terrace of the village; it is extremely local, as that and Weybridge, in Surrey,

T

are the only two spots where I have known it to be found. It is one of the most vivid fliers among the bees, and darts about, especially during brilliant sunshine in June, with the velocity of a sand-martin, and its note is shrill, but harmonious; it is not often caught upon flowers, being so extremely alert, but has been seen to visit the common Viper's Bugloss (*Echium vulgare*). The *M. odontura*, the last of the second division, which is known only in a single male specimen in the cabinets of the British Museum, is one of Dr. Leach's west country captures, of which nothing precise is known, and it is only noticed here on account of the singular peculiarity of the armature of the apex of its abdomen, which brings it closer to the genus *Osmia* in that particular, although the majority of the males of the genus have the terminal segment slightly furcated.

In these observations I have commenced with the division which contains the type, and to which the present name of the genus would attach from that circumstance, were it ever thought desirable to separate those species, which have dilated anterior tarsi in the males, into a distinct genus, but which I could scarcely recommend. In the arrangement of the species in the preceding list, I have placed these latter first, from their more symmetrical appearance in the cabinet, by leading down to the terminal smaller species in due order, from these larger and more conspicuous ones.

The *M. Willughbiella* and *maritima* prefer decaying wood, and they have been found upon decaying Willows in the Midland Counties in extreme abundance; they might be called gregarious were the material within which they burrow connected in a continuous plane. The *M. Willughbiella* makes use of the leaves of the

Rose and of the Laburnum, but the *M. maritima* seems to prefer the leaves of the Sallow. The *M. circumcincta* invariably burrows in banks, confirming the semi-gregarious habits of the genus, where it forms large colonies, and it is only by accident that it constructs secluded and solitary nests; it also makes use of rose leaves for lining its apartments. The insects are subject to the molestation of bee-parasites of the genus *Cœliorys*, the *C. quadridentata* having been bred from the cells of this latter species,—that parasite also frequenting the *M. Willughbiella*, and the *C. vectis* is well known to infest the *M. maritima*. Thus, it appears to be only the species of this division with the dilated tarsi that are exposed to such incursions, there being no record of parasites frequenting the division in which the males have simple anterior tarsi. Besides this bee-parasite, they are also subject to the attacks of some dipterous insect, whose larvæ destroy the larvæ of the *Megachile*. Much difficulty exists in separating the females of some of the species from each other; in others the specific character is sufficiently noticeable. It is a singular concomitant that those males with the dilated anterior tarsi have the apical joint of the flagellum of the antennæ considerably compressed and also dilated laterally.

The proceedings of these bees are very curious. Although the tubes they usually form are long, they are so constructed as not to branch far away from the exterior of the material into which they bore,—sound or putrescent wood or earth, or old mortar joining the bricks of walls,— if in the second material, they usually follow the putrescent vein, and their tunnel in every case is rarely further than an inch or an inch and a half from the external surface. Both the sides of the tube, and the cells they form within them, will necessarily vary in diameter and length with the size of the species, but in the larger species they are about an inch and a quarter long and half an inch in diameter. Some entomologists have surmised that different species use the leaves of different plants for lining their cells; this, however, is not strictly the case, as shown in the preceding remarks; but, although not so, the series of nests in the same tube are always lined with cuttings from the same plant; perhaps a varying caprice operates upon each day's labours and changes the plant, influenced by the drift of the wind or some casual freak.

The cylindrical tube being prepared, which is done very similarly to the way in which it is practised by all the labouring genera, by the gradual removal of the particles of the wood, or sand, or earth of which it consists, the insect's instinct prompts it to fly forth to obtain the requisite lining, that the lateral earth may not fall in, or the wood taint the store to be accumulated for the young, for it is before this is done that the upholstery is commenced. Having fixed upon the preferred plant, Rose-bush or Laburnum or Sallow, or whatever it may be, it alights upon the leaf, and fixing itself upon the edge, it holds it with three legs on each side, then using its mandibles as the cutter of silhouettes would his scissors, and, just as rapidly as he cuts out a profile, does this ingenious little creature ply the tools it is furnished with by nature. The oval or semicircular cutting being thus speedily dispatched, with the legs still clinging to the surfaces, the insect biting its way backwards, the piece cut off necessarily remains within the clutch of the legs, and, when about falling, the rejoicing labourer expands her wings and flies off with it with a hum of delightful triumph, the cutting being carried perpendicularly to her body. In a direct line she wings her way to the receptacle, and arrived at the mouth of the aperture within which she has to convey it, she rolls it to its requisite tubular form and thrusts it forward to the bottom of the cavity. The first piece for the lining of each cell is always oval and larger in proportion than the succeeding ones, which, to the number of three or four, are semicircular, the first piece having an extra use to serve in forming a concave bottom to the cavity. Having completed the requisite manipulation, for adjusting it to shape the external lining of the bottom and sides of the first cell, she withdraws backwards, again flies off, and, as if she had traced a trail in the air, or had marked its limpidity with a frothy surge, like that left in the wake of a ship, to note the road for her return, back she wends to the same plant, and proximately to the spot of her recent triumphant exploit renews the operation, but the result of which, this time, is to be semicircular. Home she flies again, and the arrangement within of this piece is different to that of the first, for this is simply tubular, and so placed that it imbricates with its cut margin within the serrated edge of the first and the third, and in case of a fourth the fourth also is similarly placed, so that one laps within the other, the edges of two of these cuttings never being conterminous. The number of the coatings is apparently regulated by the drier or moister condition of the substance in which the tunnel is drilled. Another duty has now to be performed, indeed, that for which all the preceding labours were undertaken,—the provision for its young, wherein it perpetuates its kind,—and thus on and on flows the wonderful stream of life, whose origin who shall estimate through the millennia it has hitherto so placidly and uniformly traversed, and whose termination who shall predict? Having completed the requisite store of honey mixed with pollen, this is carried to the brush with which the under side of the abdomen is furnished, by means of the posterior legs. The honey and pollen are gathered from different kinds of thistles, whence it acquires a reddish hue and looks almost like conserve of roses, and the nest is filled with it to within a line of its top; the egg is then deposited, but the coating of leaves, which enclose the cell completely, secures the store from lateral absorption, although the mixture is rather more fluid, consisting of a relatively greater quantity of honey than is usual, excepting perhaps in the case of *Ceratina*, and although no viscous secretion is used to bind the leaves together, which retain their position from merely lateral pressure. The cell has now to be closed, and the artificer knowing that the transverse section of the cylinder is circular, again flies forth, and without compass, but with all the accuracy with which Leonardo da Vinci struck a circle with his pencil, to testify his mastery, cuts the leaf again in that form, and as surely: and, three or four, or five or six times, repeats this operation, returning each time with each piece, so many having been variously observed. The separation between the cells being thus consolidated, it is further thickened by the lateral, spare, protruding edge of the leaf first introduced lapping over it. The whole process is again renewed in the same manner as at first, the bottom edge of the cutting of the external leaf is again curved to form a concave bottom to the next cell, and the sides are similarly formed, and each cell fits the preceding like the top of one thimble placed in the mouth of another. The repetition of all this is continued until the completion of the five or six cells necessary to fill the tube, when another is formed with the same routine, if her store of eggs is not exhausted; and the orifice of the tube, upon the completion of the last cell, which is closed in the usual way, is filled up with earth. Should any casualty interfere with her labours or temporarily derange their utility, without the obstruction being one that would permanently affect it, the remarkable patience and rapidity with which the repairs are effected, or the obstructions removed, is worthy of all admiration,—the στοργή, or love of offspring, being the predominant passion which overthrows and controls every difficulty.

When full fed, the larva spins a thick cocoon of silk, which is attached to the sides of the cell; the outer coating of this cocoon is of a coarser and browner silk than the interior, which is formed of very delicate threads of a slaty-whitish colour and of a close texture, and which is as lustrous as satin. The exact period of their evolution from this state is not recorded, but it is probable that they pass the winter enveloped in their cocoon as pupæ, and in their season come forth the following year.

Genus 20. ANTHIDIUM, *Fabricius.*

(Plate XIII. fig. 1 ♂ ♀.)

APIS ** *c* 2 β, Kirby.

Gen. Char.: BODY subglabrous. HEAD transverse, as wide as the thorax; *ocelli* in a triangle on the vertex, which is flat; *antennæ* shortish, slender, filiform, subgeniculated; the *scape* stouter than the flagellum, subclavate, first joint of flagellum globose, the remainder subequal; *face* flat; *clypeus* triangular, truncated at its base, slightly rounded in front and convex; *labrum* longitudinally oblong, the sides parallel and concavo-convex; *mandibles* dilated at the apex, where they are quinquedentate; the *alternate teeth* smallest; *cibarial apparatus* long; *tongue* very long, tapering to its extremity; *paraglossæ* very short, one-sixth the length of the tongue, coadunate at the base and subhastate; *labial palpi* more than half the length of the tongue, the two first joints very long, the second the longest, and both tapering to the acute extremity of this, where, just before its apex, the third very short subclavate joint articulates with the still shorter terminal joint conterminous with it; *labium* one-third the length of the tongue, its inosculation with an acute projection in the centre; *maxillæ* as long as the tongue, subhastate and acuminate; *maxillary palpi* springing from a deep sinus at its base, very short, two-jointed, the basal joint the shortest, and the second obtuse one terminating with a few rigid setæ. THORAX subglobose; *prothorax* inconspicuous; *mesathorax* slightly convex, wing-scales large; *scutellum* lunulate, projecting and impending over the metathorax, which is truncated; *wings* with two submarginal cells, and a third indistinctly commenced, the second slightly the longest, and receiving the two recurrent nervures one at each extremity; *legs* moderate, subsetose, the tibiæ fimbriated along the edges, the anterior *spurs* slightly palmated; the *plantæ* of the four anterior pairs longer than their tibiæ, but those of the posterior not quite so long, and all densely clothed all round with a brush of short close hair; the *claws* distinctly bifid. ABDOMEN semicircular, very convex; the base truncated and hollowed to fit the metathorax; the *segments* slightly constricted, the *terminal segment* transversely concave, and its apex terminating in three slight angles; the *venter*, which is flat, is densely clothed from the second segment with parallel layers of equal, moderately long, shining hair, the segment being distinctly indicated by these layers.

The MALE differs in being considerably larger; the *mandibles* merely tridentate; the *legs* longer and more robust; the *tibiæ* and *tarsi* more densely fimbriated externally, and the tarsi relatively much longer; the *abdomen* densely edged laterally with short curled hair, the terminal segment with three processes, the lateral ones strong and curved internally, the central one shorter and straight, and the penultimate segment transversely concave, with a strong tooth on each side curved externally, and the venter glabrous beneath.

NATIVE SPECIES.

1. *manicatum*, Linnæus. 5-8 lines. (Plate XIII. fig. 1 ♂ ♀.)
manicatum, Kirby.

GENERAL OBSERVATIONS.

The generic name in this instance seems to be manufactured from the root ἄνθος, *a flower*. I cannot trace any other derivation as it may not be attributed merely to the habits of the species in frequenting flowers, for is not this the prime function of all the bees, wherein they fulfil a most important office in the economy of nature? How easy might it have been to regulate that flowers should fertilize themselves, as many do without any extraneous intervention, but by this wise and benevolent ordination a tribe of sensitive creatures is introduced to be perpetuated by the perpetuation they supply to that which supports them, and in this circle of reciprocal good offices lend an additional charm to the genial seasons, by the animation which they give to the face of nature, in embellishing the plants they visit with their vivacity and music.

These bees are gay insects, for both sexes are richly spotted with yellow, and they present the single instance which occurs amongst our bees of the male being considerably the largest, and so boisterous is he in his amours that he forcibly conveys his partner to the upper regions of the air, where she is compelled to yield to his solicitations. His whole structure is fully adapted to carry out this violent abduction, as well in the length and power of his limbs as in the prehensile teeth with which the apex of his abdomen is armed.

We have but one species of the genus, although the southern parts of the Continent abound in them. The habits of ours differ very considerably from those of the preceding genus. First, in the peculiarity just described, and then in the formation of their nests. They do not, like the majority of the wild bees, excavate or bore a cavity for themselves, but take one already formed by the xylophagous larva of some considerable insect, such as *Cerambyx moschatus*, or *Cossus ligniperda*. This they line, to the depth suitable to them, with cottony down which they scrape from the leaves or stalk of the Woolly Hedge-nettle (*Stachys Germanica*), the Wild Lychnis (*Agrostemma*), and other woolly-leaved plants. In collecting this wool the insect is very active, scraping it off rapidly with its broad mandibles, and as this is doing she gradually rolls it up into a little ball, making with the vibration of her wings a considerable hum all the time she is gathering it, and when the ball is sufficiently large she flies off with it to her nidus; this operation she continues until sufficient is accumulated for her purpose, which consists in lining the cavity with the material; she then forms cells within it in succession, gluing the same material together to resist the escape of the mixed store of pollen and honey she intends to fill it with, having in the operation smoothed the sides of the cell which is closed after the deposit of the egg, and another similar cell is then proceeded with, and this is repeated until the selected cavity is filled, or that she has exhausted her store. Having completed her labours, she wanders away. Sometimes the cavity is large and admits of the conjunction of many of these cells together; in that case they are all collectively covered with the same envelope of downy substance. The larva, having consumed its entire store of food, spins a cocoon of brown silk wherein it remains throughout the winter, and with the evolution of spring, feeling its propulsive energy, it changes into the pupa. In June and July, but earlier if the weather be continuously warm, the imago comes forth in its maturity, to live its little life of labour intermingled with pleasure, and in its pleasing hum to give cheerful notification of its perfect satisfaction.

Genus 21. CHELOSTOMA, *Latreille.*

(Plate XIII. fig. 2, ♂ ♀.)

APIS ** *c* 2 γ partly, Kirby.

Gen. Char.: BODY nearly glabrous and coarsely punctured. HEAD subglobose, rather wider than the thorax;

ocelli in a triangle in the centre of the vertex, which is broad and slightly convex; *antennæ* short, subclavate, geniculated, the scape nearly one-half the length of the flagellum and more robust; the first and second joint of the flagellum subclavate, the basal one the longest and most robust, the remainder short, subequal, and gradually enlarging to the apical one, which is obtuse and as long as the basal joint; *face* flat, slightly convex between the insertion of the antennæ; *cheeks* large and protuberant; *clypeus* concave, projecting, lobated in front, where it is slightly emarginate in the centre; *labrum* elongate at its articulation, broader than beyond, and from this expansion immediately and abruptly contracting, from the inner angles of the contraction waving to about three-fourths its length, whence it is produced into an equal truncated oblong; *mandibles* bidentate, external tooth acute, inner one obtuse; *cibarial apparatus* long; the *tongue* twice the length of the labium, narrowest at its base and obtuse at the extremity, and clothed with short setæ; *paraglossæ* very short, coadunate at the base and acuminate; *labial palpi* two-thirds the length of the tongue, with the three first joints membranous and flat, conterminous and tapering to their extremity, the first joint about one-half the length of the second, the third twice the length of the fourth, which is clavate and articulated within the apex of the third; *maxillæ* subhastate and acuminate, as long as the tongue; *maxillary palpi* very short, rather stout, the joints subequal and the terminal one acute. Thorax oval, convex; *prothorax* inconspicuous; *wing-scales* rather large; *scutellum* transversely quadrate, convex; *post-scutellum* transverse, linear; *metathorax* gradually declining, with a glabrous triangular space at its base;

wings with two submarginal cells nearly equal and a third commenced; the second receives both the recurrent nervures, the first beyond its commencement and the second before its termination; *legs* shortish, subsetose, the anterior spurs short, broad, and emarginate at the apex; the *posterior plantæ* with a compact dense brush within; *claw-joint* long; *claws* simple. ABDOMEN longer than head and thorax, subclavate, convex above, retuse at the base, and the apical segment obtuse at its extremity, the venter flat, its segments clothed from the second with dense parallel brushes of longish hair for the conveyance of pollen.

The MALE differs in having the *head* less conspicuously globose; the *cheeks* less protuberant; the whole body more pilose, the anterior *spurs* robust, short, and abruptly obliquely truncated; the *antennæ* slender, filiform, much longer than in the female, but not much longer than the head, and from the fourth to the ninth joints serratulate within, adapting it to a sharp curve; the *abdomen* being equal, cylindrical, retuse at its base, convex above, and flat on the venter, where it has a longitudinal deeply concave mucro in the centre of the second segment, which concavity runs along all the subsequent segments, and it is densely pilose on the fourth; the terminal dorsal segment being deeply emarginate in the centre and produced on each side into a broad obtuse process; the *claws* are more robust than in the female and bidentate; the posterior pair being subclavate, and their single tooth abruptly reflected.

NATIVE SPECIES.

1. *florisomne*, Linnæus, ♂ ♀. 3–5 lines. (Plate XIII. fig. 2 ♂ ♀.)

maxillosa, Linnæus.
maxillosa, Kirby.
2. *campanularum*, Kirby, ♂ ♀.   2-2½ lines.

GENERAL OBSERVATIONS.

These insects are named from χηλὴ, *a forceps*, and στόμα, *a mouth*,—in allusion to the forcipate form of the mandibles, which are strong, and cross each other in inaction.

They and the next genus are styled *carpenter bees*, but they are not more consistently thus called than might be *Anthophora furcata* and the genus *Ceratina;* they, in fact, like the latter, just as often avail themselves of an empty straw to form their cells in, or the cylinder that has been drilled by some xylophagous beetle of their own size, as they themselves drill into palings and solid wood for the purpose, but when they do this, it is facilitated to them by their powerful mandibles and their square and strong head. They are certainly very compactly formed, their structure being indicative of great power, of course relatively to their size. When they drill their cylinders themselves they are extremely persevering in its execution, and in the process, the material they extract, which is like fine sawdust, they withdraw from the depth of the cavity by passing it beneath them, and pushing it out at the orifice by means of their posterior legs and the apex of the abdomen, for they are too long to be able to turn within the cavity they have formed, its capacity not being sufficient to permit this, as it is very little larger in diameter than themselves. I have repeatedly watched them in these operations.

Having found or drilled a suitable cylindrical tube, they

do nothing further to it but collect a sufficient store of provender for the nutriment of the young one, upon which they deposit the egg which is to produce it. The insect then flies away to collect a small quantity of clay intermingled with sand, and this they knead together by means of a viscous secretion which they disgorge, and this forms a concrete that hardens firmly and rapidly; to anticipate its rapid drying they speedily fly back, carrying this small ball within their mandibles, and with it they cover over the provision they have collected, and which, adhering to the sides of the cavity, forms a firm and hard division, effectually separating it from the next store of provision that is to be accumulated for the supply of the larva that will be hatched from the egg that is to be deposited, and the same process is repeated again and again until all the eggs are laid. In their development, which takes place near midsummer, the males precede the females by about ten days. They associate sometimes in colonies, often using the tubes of the straw thatch which covers cottages for their nidus.

These bees are subject to the parasitical intrusion of *Fœnus jaculator* and *assectator*, which I have repeatedly caught at Battersea, hovering opposite the cells of these insects bored in the shingles forming the enclosure of an old garden outhouse. These parasites are themselves peculiar creatures, forming a type distinct from the Ichneumons, and belonging to the group *Aulacus*, upon which see my paper in the 'Entomologist,' June, 1841. In these insects, the abdomen springs from immediately beneath the scutellum. *Chrysis cyanea* and *ignita* are also bred at the expense of these bees, neither of the species of which are uncommon; the smaller one, the *C. campanularum*, which is the smallest of our true bees, excepting perhaps one or two of the *Nomadæ*, I used to find in abundance upon the railings of the fields that skirt Hampstead Heath, on the right-hand going from London, parallel with the Vale of Health, and thence rising to the Holly enclosure of the Earl of Mansfield's mansion. This spot has been productive to me of many very choice aculeate *Hymenoptera*, and supplied me with them in abundance at a time when even the chief metropolitan collections were bare of them. It has also furnished me with several very desirable *Diptera* of extremely rare genera. The male of the larger species of this genus Linnæus called *florisomne*, from its habit of curling up its abdomen and antennæ, and passing the night in flowers. Those which they chiefly frequent are the species of Wallflower, and the *Campanula*, especially the round-leaved Throatwort.

Genus 22. HERIADES, *Spinola*.

(Plate XIII. fig. 3 ♂ ♀.)

Apis ** *c* 2 γ partly, Kirby.

Gen. Char.: BODY glabrous and much punctured. HEAD globose and curving to the thorax posteriorly; *ocelli* in a triangle far forward on the vertex; *antennæ* slightly subclavate, the scape not half so long as the flagellum, the first joint of which is robust, subclavate, and twice the length of the second, which, with the rest, are subequal, very slightly lengthening to the terminal one, which is as long as the basal one and laterally compressed; *face* slightly convex, cheeks large and convex; *clypeus* lunulate, convex, and with two minute central teeth on its front margin; *labrum* longitudinally oblong, rather broadest at the base and slightly waved laterally, concavo-convex and subemarginate at the apex; *mandibles* subequal, tridentate at the apex, and the central tooth obtuse; *cibarial apparatus* moderately long, *tongue* twice the length of the labium, with a small knob at its apex; *paraglossæ* very short, almost obsolete, coadunate at the base; *labial palpi* two-thirds the length of the tongue, the two first joints membranous and long, the first one-third the length of the second, which tapers to its acute extremity, before the end of which the two terminal, subclavate, very short, subequal joints are inserted; *labium* half the length of the tongue, slightly produced in the centre of its inosculation; *maxillæ* subhastate, two-thirds the length of the tongue; *maxillary palpi* three-jointed, short, robust, equal, and collectively subfusiform, the terminal one rather acute. THORAX globose; *prothorax* inconspicuous; *scutellum* lunulate; *post-scutellum* linear, transverse; *metathorax* declining; *wings* with two submarginal cells, and the commencement of a third indicated, the second larger than the first, subtriangular, and receiving both the recurrent nervures, one at each of its extremities; *legs* short, rather robust, subsetose and spinulose; *posterior tibiæ* convex externally and with their plantæ rugose, the latter covered beneath with a dense brush of short hair; *claws* simple. ABDOMEN cylindrical, convex above, retuse at the base, and the first and second segments slightly constricted at their extremity, obtuse, and from the end of the third segment sensibly declining to the apex; plane on the venter, where, from the second segment, the plate of each, excepting the glabrous terminal one, is covered with a dense brush of short hair for the conveyance of pollen.

The MALE differs in the *antennæ* being rather longer, more distinctly filiform, the seventh segment of the *abdomen* concealed under the extremity of the sixth, and the *venter* from the third segment longitudinally deeply concave, the plate of the third itself covered with hair; the *claws* more robust and each equally bifid, not bidentate.

NATIVE SPECIES.

1. *truncorum*, Linnæus, ♂ ♀. 3–3½ lines. (Plate XIII. fig. 3 ♂ ♀.)
truncorum, Kirby.

GENERAL OBSERVATIONS.

The names of insects are not always very aptly given, for the only available derivation of this appears to be from ἔριον, *wool*; in allusion to the clothing of its venter; but, if so, it should be spelt without the H, for the first letter is without an aspiration. The habits of these closely resemble those of the preceding genus, to which they have a great personal likeness, and therefore their natural history would be but its reiteration. Our solitary species is a rare insect, but I expect western England would produce it. It is like those of the preceding genus, of a uniform black colour, punctured, but it approximates more closely than they do to the type of form exhibited in the genus *Osmia*. They visit the same flowers as the preceding genus.

Genus 23. ANTHOCOPA, *St. Fargeau.*
(Plate XIV. fig. 2 ♂ ♀.)

Gen. Char.: BODY glabrous, subpubescent, shining. HEAD subglobose, as wide as the thorax; *ocelli* placed in a slight curve on the summit of the vertex; *antennæ* short, geniculated, the *flagellum* subclavate seen in front, but seen from above, owing to the compression of the terminal joint, subfusiform, the first joint of the flagellum globose, rather robust, the second short, subclavate and subequal with the rest, which increase gradually in length and substance to the terminal one, which is the longest, and laterally compressed; *face* flattish; *clypeus* subquadrate, very convex and very pubescent; "*labrum* oblong, quadrate; *mandibles* strong, tridentate; *labium* (tongue) long, filiform; *labial palpi* having the third joint articulated externally on the outer side of the second; *maxillary palpi* four-jointed." Thorax globose; *scutellum* lunate; *post-scutellum* transverse, linear; *metathorax* rounded; *wings* with two submarginal cells and the commencement of a third just indicated, the second very slightly larger than the first, and receiving both the recurrent nervures, the first just beyond its commencement and the second close to its termination; *legs* short, rather robust, subsetose; the *posterior tibiæ* externally convex and the *posterior plantæ* with a dense, short brush beneath; the *claws* simple. Abdomen cylindrical, retuse at the base, convex above, declining from the base of the fourth segment to the extremity, the first and second segments very slightly constricted, the margin of the posterior one, at the apex, slightly crenulated, the ventral segments plane and from the second covered with a dense brush of parallel hair, excepting the sixth, which is reflected laterally and longitudinally, convex down the centre.

The male differs in having "the sixth segment of the *abdomen* emarginate, and with a strong tooth on each side; the terminal segment emarginate, thus producing two strong, lateral, obtuse teeth, the ventral plates of these same segments emarginate at the extremity, and the emargination fringed with hair; the claws bifid."

NATIVE SPECIES.

1. *papaveris*, Latreille. (Plate XIV. fig. 2 ♂ ♀.)

GENERAL OBSERVATIONS.

Named by St. Fargeau from ἄνθος, a *flower*, and κοπὴ, a *cutting* or *incision*, from its habit of cutting sections out of the petals of the common scarlet poppy with which to line the cells it forms within the cylinder it excavates, just as *Megachile* does with the leaves of various plants. It is noticed as British upon the faith of the specimens introduced by Leach into the cabinets of the British Museum and presumptively caught in the west or south-west of England, a region rich in rarities. Rennie in fact tells us that he has found it at Largs, in Scotland. One of Leach's specimens I received in exchange from that establishment in 1842, and which is now in the possession of Mr. Desvignes, to whom my collections passed in the following year. This genus forms a sort of combination between the genera *Megachile* and *Osmia*, it having the upholstering habits of the former in the mode with which it lines its nest, and the general habit of the latter. At a first glance, before its habits were known or its structure examined, even an experienced entomologist might have placed it under *Osmia*, as an unrecognised species, for it very strongly resembles the *Osmia leucomelana*. This proves how very inconclusive habit is as an index to habits, the latter of these insects drilling into the pith of brambles, and the *Anthocopa* tunnelling cylinders into the hardest trodden roads or pathways and lining them with its crimson hangings.

From the extreme rarity of the insect, I have been unable to examine the cibarial apparatus, and thence to ascertain upon what substantial grounds the generic distinctions are based, which separate it from *Osmia*. Whether it was these mere habits of the insect which induced Le Pelletier de St. Fargeau to establish the genus I do not know, but he is always extremely slovenly, and therefore very unsatisfactory in his characteristics, which are never framed in a strictly explicit manner. In consequence of all these difficulties, I have merely been able under the generic character to introduce such as he has given, which I could not derive from the personal external inspection of Mr. Desvignes' female (my own selection of whose bees for the purposes of this work he has been so kind as to lend me, and whom I thus publicly present with my best thanks). I have therefore compounded a character as well as I could from St. Fargeau's descriptions, inserted in the tenth volume of the '*Encyclopédie Méthodique*,' and from his work on the *Hymenoptera*, forming one of the '*Suites à Buffon*.'

The habits of these bees, as said above, are to excavate vertical cylinders in hard down-trodden pathways and roads, by the sides of fields where corn is grown, and where consequently the common red poppy is abundant. From the petals of the flowers of this plant they cut out semicircular pieces, precisely as is done by *Megachile* with the more rigid leaves of shrubs and trees, and convey them home and line their nests with them, just as is practised by that genus with those leaves.— with this difference merely, that a sufficient portion of the upper edge of the pieces of the petals used is left projecting, for the purpose of forming a covercle to the nidus, and which, when filled with provender and the egg deposited, is refolded over it and covered in, and it is closed up with earth. They then proceed to make another excavation, which is treated in the same manner, for they deposit only one larva in a tube. If disturbed in their retreat, they will show themselves at its mouth, like *Dasypoda*, to see what is the matter.

I would urge our collecting entomologists, especially those who have the opportunity of hunting up the west of England, to use due diligence and strive to confirm the native existence of this bee and add specimens to the cabinets of their fellow-entomologists.

Genus 24. Osmia, *Latreille*.
(Plate XIV. figs. 1 and 3 ♂ ♀.)
Apis ** c 2 δ, Kirby.

Gen. Char.: Head subglobose, concave, posteriorly fitting the prothorax and about as wide as the thorax; *ocelli* placed far forward on the vertex, which is wide and convex, in a curved line; *antennæ* filiform, sometimes subclavate, short, and geniculated, the *scape* robust, as long as the four following joints, the basal joint of the *flagellum* globose, its second joint clavate and as long as the terminal one, the remainder short, subequal, and gradually but slightly increasing in length; the *face* flattish; the *clypeus* a truncated triangle, convex; *labrum* longitudinally oblong, a little laterally distended at the articulation, from whence the sides are parallel; *mandibles* broad at the apex, obscurely tridentate, the internal teeth obtuse and short; *cibarial apparatus* long; the *tongue* three times the length of the labium, clothed with short hair and tapering from the base to the acute apex; *paraglossæ* very short, coadunate at the base and acuminate at the apex; *labial palpi* more than half the length of the tongue, the two first joints membranous and long, the basal one the broadest, seated on a petiole and not so long as the second, which tapers to an acute point, before the apex of which the remaining two short subclavate conterminous joints articulate; *labium* about one-third the length of the tongue, acutely produced in the centre of its inosculation; *maxillæ* as long as the tongue, subhastate and acuminate; *maxillary palpi* four-jointed, rather short, the joints subequal and subclavate, but the second is both the most robust and slightly the longest. THORAX oval or globose; *prothorax* inconspicuous; *scutellum* lunulate and convex; *post-scutellum* transverse and linear; the *metathorax* abruptly truncated; *wings* with two submarginal cells, and a third distinctly commenced, the second the longest, and receiving both the recurrent nervures, the first towards its centre and the second near its termination; *legs* moderate, setose, the plantæ of all with a dense brush beneath; *claw-joint* longer than the three preceding; *claws* simple. ABDOMEN short, cylindrical, convex, the terminal segment slightly pointed, the ventral segments densely pilose in parallel lines from the second.

The MALE differs in having the *antennæ* longer and always filiform, the ventral segments very concave, and the terminal dorsal segment variously mucronated, tuberculated, spinose or serrated, and the claws bifid.

NATIVE SPECIES.

1. *leucomelana*, Kirby, ♂ ♀. 3–1½ lines. (Plate XIV. fig. 3 ♂ ♀.)
2. *spinulosa*, Kirby, ♂ ♀. 3–4 lines.

3. *pilicornis*, Bainbridge, MS., ♂ ♀. 4–4½ lines.
4. *bicolor*, Schrank, ♂ ♀. 4–5 lines. (Plate XIV. fig. 1 ♂ ♀.)
5. *fulviventris*, Panzer, ♂ ♀. 4–5 lines.
   Leaiana, Kirby.
6. *ænea*, Linnæus, ♂ ♀. 3–4½ lines.
   cærulescens, Linnæus, ♀.
   cærulescens, Kirby, ♀.
7. *parietina*, Curtis, [V. 222.] ♂ ♀. 3–4 lines.
8. *xanthomelana*, Kirby, ♂ ♀. 4–7 lines.
   atricapilla, Curtis, [V. 222.] ♀.
9. *aurulenta*, Panzer, ♂ ♀. 4–6 lines.
   tunensis, Kirby.
10. *rufa*, Linnæus, ♂ ♀. 3–6 lines.
    bicornis, Linnæus.
    bicornis, Kirby.

GENERAL OBSERVATIONS.

Named from ὀσμή, *sweet-scent*, from some fancied idea of their possessing the property of emitting a sweet odour; but this, although it is the case with many of the bees,—for instance, with the genera *Prosopis*, *Halictus*, *Nomada*, some of the *Anthophoræ*, *Saropoda*, and the male *Bombi* and *Apathi*,—I have not noticed in any of this subsection, the *Dasygasters*, and therefore not in any of the present genus. It is possible that when richly laden with pollen, this may emit some smell, but I am not aware that any of the scent of flowers lies in the anthers or their pollen, although this in some cases has a spermatic odour pointing to its express function; but be this as it may, such is their name. These as a group are what are called the 'Mason Bees,' from the habit they have of agglutinating particles of sand or earth mixed with minute pebbles, scarcely larger than grains of sand, or raspings of wood combined in the same manner, with a secretion which they emit, and of which they form their cells. The instinct of the creature prompts it to be speedy in the operation, as the material, like plaster of Paris, dries very rapidly to a hard substance. Whether they have the power of softening the edges as the manufacture of the cell proceeds is not known, nor whether, as they add the material, it instantaneously consolidates itself, but the colour of the structures themselves would indicate a simultaneous mixture. This could not be the case, if the mortar or mixture were formed away from the domicile and brought home in little pellets, each being added upon the insects' arrival, although they obtain it all from the same spot, whence arises its uniformity in colour, and they are speedy in the formation of their nests. These cells are rather rough externally, according to the nature of the material of which they are composed, but they are very smooth within. The nature of the cells varies with the places of their deposit, which is dependent upon the idiosyncrasy of the species. Thus, those which construct their cells in wood, form them of moistened particles of wood, and those which make them in cavities of any kind, in the earth, beneath stones, or within empty snail-shells, make a mortar of earth and sand and small pebbles. Some are strictly uniform in the selection of the material wherein they build, but others are perfectly indifferent to its locality, and adopt either earth or wood, and sometimes the mortar of walls, sandbanks or chalk cliffs. According to the nature or the size of the receptacle which they select, is the adjustment of these cells. Where the cavity is restricted they place them end to end, but where it is more roomy they affix them side to side, completely adapting themselves to the circumstances of the locality as I shall instance below, in the description of the special habits of the more conspicuous species. I have elsewhere referred to the metallic colouring of many of the species of this genus, and amongst them is found the greatest sexual disparity of personal appearance, the *O. leucomelana*, and one or two of the neighbouring species being, perhaps, the only ones wherein uniformity of appearance would unite the partners together. The majority are very pubescent insects, and the females of the terminal species in the foregoing list are remarkable for a couple of inwardly curved horns, springing from the base of the clypeus just below the insertion of the antennæ, an appendage usually a male attribute.

There is very great dissimilarity in the habits of the various species, whence no single characteristic will embrace them, nor is there any distinctive feature whereby the genus might bear subdivision, either from habits or habit, as will be collected from the following cursory survey of their special natural history.

Thus the first species, the *O. leucomelana*, named so from the white decumbent down which edges the black segments of the abdomen, extracts the pith from bramble-sticks, and its cells are formed and closed with a composition made of triturated wood or leaves. The cylinders it forms are usually about five inches deep, and within this it constructs about the same number of cells proportionate to the small size of the insect. These are midsummer insects, coming forth in June and July; they are very local, but seem to abound in the vicinity Bristol, whence Mr. Thwaites formerly sent me speci-

mens. A very few days serve for the hatching of the
larva, which spins a slight silken cocoon, and in this
dormitory it reposes until its season again comes round.
Under the influence of the following first genial spring
weather, the larva is transmuted into the pupa, and the
active little imago comes forth upon the settlement of
our variable spring, in the merry days of June, and thus
is perpetuated the circle of its existence, but which is
sometimes abridged by its special parasite, the pretty
little *Stelis octomaculata*. Many of the species in the
males are distinguished by a peculiar armature of the
apex of the abdomen; the second being named by Kirby
from the circumstance. A very remarkable singularity
distinguishes the males of the third species, in the fringe
of short hair that runs along the flagellum of its antennæ. This, I believe, was first noticed by the late
Mr. Bainbridge, a very active practical entomologist,
who took the insect at Darenth or Birchwood, and distributed specimens with this manuscript name attached,
which has since been appropriated by another entomologist to whom the science was wholly unknown at that
time, but as it is scarcely consistent with scientific
courtesy to adopt such a course, and as the MS. names
of Linnæus and Kirby have been retained, where it
was authorized by their being attached to undescribed
species, I have restored to Mr. Bainbridge his just
rights, and have claimed the same for myself, in the
case of *Andrena longipes*, and which many cabinets must
still possess with my name attached, in my own writing,
unless their possessors have chosen to adopt the illegiti-
mate parentage; for the entomologists of my own stand-
ing well know that I always freely distributed specimens to all who desired them of the many very desirable

insects which I have captured in the course of my entomological career. The fourth and the ninth species, the *O. bicolor* and *O. aurulenta*, have very much the same habits, both usually burrowing in sandbanks, sometimes however in wood, in which case the perforation, contrary to the mode of wood-drilling bees, is made upwards, a sagacity or instinct which saves it much trouble, for the particles as they are removed by the mandibles are passed beneath the insect, and their own gravity carries them downwards, and thus the insect saves itself the labour of conveying them out as they accumulate in inconvenient quantities. The cells in this case are placed end to end. When they burrow in the earth, the latter species often associate gregariously in large numbers, and if they select a cavity, instead of tunnelling it themselves, and it be too large to take one cell upon the others, they form them side by side, and thus fill the space. This is the case when they adopt snail-shells as the receptacle for their incunabula, and this is done by both these species, and the shells they select are the empty ones of *Helix nemoralis, hortensis*, and *adspersa*. The capacity of the latter shell being much greater than that of the others, and too wide for a single succession, she fills the interval by placing them side by side, and with the increase of the whorl of the shell towards its orifice she places them across the space, and thus completes her task. In the former shells, the cavity at first admits of the succession of but one upon the other, but with its enlargement she places them side by side, and this repeated fills the hollow. Its aperture is then closed with earth and pebbles or sticks agglutinated together, as described at the commencement. The *O. fulviventris* burrows in wood, and upon this species the *Stelis phæoptera* is parasitical; and that very pretty but extremely common species the *O. ænea*, in which the male is of a rich bronzy tint, and the female of a beautiful blue, verging sometimes to nearly black, burrows also in wood, although sometimes it capriciously selects old walls or chalk-cliffs, and is subject to the incursions of the same parasite. Perhaps the most extraordinary species is the *O. parietina*, figured and named by Curtis, and which he first found at Ambleside; it has since been found in the Grampians very considerably above the level of the sea, and it is thus essentially a northern species both from altitude and locality. It would appear that this species selects some flat stone of about a foot in surface, lying upon the ground over a hollow spot. Such a specimen, sent to the British Museum, had attached to its under side two hundred and thirty cocoons, indicative of a considerable colony, or perhaps the accumulation of successive years, as one-third of these cocoons were empty of tenants. These, in their new depository, continued developing themselves in the perfect state between March and June, males appearing first. When the transformations of the season ceased, five-and-thirty were still left to present themselves another year, and the following spring these were developed; thus, including those which had already escaped when the stone and its treasure was secured, three successive seasons were occupied in their transmutations. It may be a species that requires three years for its metamorphosis, and the whole deposit of cocoons may have been the result of three years' accumulative structure, the vital activity of their northern life being perhaps more sluggish than in species frequenting the south. The last species the *O. rufa*, that in which the female is remarkable for its inverted horns, which must be for some use in its economy, is perhaps the most common of all. I have found it in abundance upon old walls with a sunny aspect at Erith, and throughout the pleasant Crays of Kent. It is indifferent as to the choice of its domicile, selecting either walls, where I have chiefly found them, sandbanks, or the decaying stumps of pollard-willows. Its processes are similar to those of some of the earlier described, but its larva is longer in full feeding, which, when it has consumed all its provender spins a tough cocoon of brown silk, wherein it undergoes its changes; some, depending much upon locality, pass into pupæ in the autumn, others hibernate as larvæ which are subject to destruction from the attacks of the *Chalcideous* insect, *Monodontomerus dentipes*, previously noticed under *Anthophora*. Some of the *Chrysididæ* also infest several of the species of this genus, and I have no doubt that *Stelis aterrima* is parasitical upon one of them, although it has not been recorded. The various species frequent many flowers, especially those abundant in the locality they inhabit, but the *O. pilicornis* chiefly affects the common Bugle (*Ajuga reptans*), and they much frequent composite flowers, especially the species of the genus *Hieracium*.

Section 2. *Cenobites* (*dwellers in community*).
Subsection 1. SPURRED.
† *Parasitical.*

Genus 25. APATHUS, *Newman.*

(Plate XV. figs. 1 and 2.)

APIS ** *e* 2 partly, Kirby.—PSITHYRUS, St. Fargeau.

Gen. Char.: BODY subhirsute. HEAD subglobose;

vertex broad, glabrous, with a deeply impressed cross upon its summit, in the centre of which the ocelli are placed in an almost straight line and contiguously; *antennæ* short, filiform, geniculated, the *scape* slightly curved, the basal joint of the *flagellum* subglobose, its second joint as long as the terminal one and subclavate, the rest short, subequal, but gradually increasing in length to the terminal one, which is laterally compressed; the *face* flat; *clypeus* transversely lunate but straight in front; *labrum* lunulate, tuberculated laterally; *mandibles* broad and obscurely bidentate; *cibarial apparatus* moderate; *tongue* twice the length of the labium, tapering from base to apex, where it terminates in a small knob, and is clothed with short hair; *paraglossæ* obsolete; *labial palpi* as long as the tongue, the two first joints long and membranous and tapering to the apex of the second, which is acute, and about one-fourth the length of the first, it has the two very short, subclavate, terminal joints, which are conterminous, and articulated just before its acute apex; *maxillæ* subhastate and acuminate; *maxillary palpi* very short, linear, and equal. THORAX globose, pubescent, concealing its divisions; *metathorax* truncated; *wings* with three submarginal cells nearly equal, or the third the largest, the second receiving the first recurrent nervure at about one-third its length, and the second is received by the third submarginal cell near its extremity; *legs* setose; the *posterior tibiæ* convex, very slightly enlarging from base to apex, rounded at the extremity externally, and unfurnished with means to convey pollen; *posterior plantæ* oblong, narrowly equal, and not auriculated; *claws* bifid. ABDOMEN ovate, convex above, deflecting toward its extremity, and subglabrous on the disk, the terminal dorsal segment triangular, and its ventral plate straight at its apex with the lateral angles reflected, making it concave beneath and subcarinated longitudinally in the centre, or also triangular and the sides of the prominent angle deflected.

The MALE differs in having the antennæ slightly longer, in being rather more pubescent, more highly and rather differently coloured, and its terminal segment merely rounded.

NATIVE SPECIES.

3. *campestris*, Panzer, ♂ ♀. 6–9 lines. (Plate XV. fig. 2. The fig. marked ♂ by mistake for ♀.)
campestris, Kirby, ♀.
Rossiella, Kirby, ♂.
Leeana, Kirby.
Franciscana, Kirby.
subterranea, Kirby.
2. *Barbutellus*, Kirby, ♂ ♀. 6–9 lines.
3. *vestalis*, Fourcroy, ♂ ♀. 6–10 lines. (Plate XV. fig. 2 ♀.)
vestalis, Kirby, ♀.
4. *rupestris*, Fabricius, ♂ ♀. 6–10 lines. (Plate XV. fig. 1 ♂ ♀.)
albinella, Kirby, ♂.

GENERAL OBSERVATIONS.

Named from *a*, privative, πάθος, *affection;* that is to say, without affection, from their habit of leaving their young to be nurtured by others, in allusion to their parasitical instincts, for the young of these bees are brought up in the nests of the *Bombi*. They form the only instance in bee-parasitism of the parasite closely, or nearly so, resembling its sitos, if not always in colour, certainly in habit. Having no labours to undergo they consist of merely males and females, but the latter, although very like the large female *Bombi*, are much less pubescent than these, for they have a broad disk, upon the upper surface of the abdomen, always smooth and shining. Both sexes appear to have free in- and egress to the nests of those *Bombi* which they infest, without any let or hindrance on the part of the latter, with whom they seem to dwell in perfect amity. In the times of their appearance they closely resemble the *Halicti* and the neighbouring *Bombi*. Thus the females, after impregnation in the autumn, having hibernated during the winter in selected receptacles, come out with the first gleams of spring conjunctively with the large maternal *Bombi*, in whose nests they have taken their long repose in perfect torpidity; and as soon as these begin to accumulate the masses of conglomerated honey and pollen whereon to deposit their eggs, the parasite takes advantage of it, lays her eggs too, and thus secures food for her offspring. There being two broods of them in the year, many are gradually developed with the advance of summer, but the great hatching takes place in the autumn, when the thistles are in blossom. Then both males and females come forth in abundance, the latter are made fertile, and their partners enjoy the brief interval of the still blossoming flowers until the usual period is put to their existence by natural decay, the first frosts, or the rapacity of insectivorous birds. Connected with this last circumstance I have a personal experience to record, and which its repetition would indicate as being one of Nature's prompting acts. A lofty sandy level, very near the high-road which leads at the upper part of Hampstead Heath, to Highgate, from which road it was separated by merely a band of whins and coarse grass, used to be a very favourite collecting place of mine, for there, and in its immediate vicinity, I have often caught, within a very brief period, more than half the genera, and a very large number of the species of the fossorial *Hymenoptera*. One particular little spot was inhabited by *Psen equestris*, rare everywhere else, and our largest *Cerceris*, who carried on their instinctive pursuits during all the summer months, but at a particular time in the autumn, varying slightly with the nature of the season, a flock of wagtails (*Motacilla*) would alight and make brief work of those fossores which were still aflight; and this was repeated season after season, as if the wagtails thought it was time that their own rapacity should stop the course of these predacious insects. But to return, the female *Apathi* then resort to the nests of the *Bombi* whence they have issued, and lay themselves up in their winter dormitory. That this must take place speedily after impregnation is rendered almost conclusive by the fine state in which their pubescence appears in the spring, which would be tarnished did they loiter about visiting flowers previous to their return home. But the labours of the female and neuter *Bombi* themselves are now over, and they would therefore find no store whereon to deposit their eggs. The parasitical allocation of these insects is as follows. *Apathus rupestris* infests *Bombus lapidarius*; *A. vestalis* the *B. terrestris*, and this forms an instance in which the parasite is not clothed in the colours of its sitos. But *A. Barbutellus* has a wide range, for it frequents the nests of *B. pratorum*, *B. Derhamellus*, and *B. Skrimshiranus*.

†† *Not parasitical. Collectors of pollen.*

‡ *Temporarily social.*

Genus 26. BOMBUS, *Latreille.*

(Plate XV. figs. 3 and 4, and Plate XVI. figs. 1, 2, 3.)

APIS ** *e* 2, Kirby.

Gen. Char. : BODY densely hirsute. HEAD small, subglobose, not so wide as the thorax; the *vertex* glabrous, with a longitudinal, short, deep channel, crossed in its centre by a deeper transverse one, wherein the *ocelli* are disposed in a very slightly curved line; *antennæ* short, geniculated, and filiform; the *scape* half as long as the flagellum, the first joint of which is globose, the second subclavate, the rest short and subequal, and the terminal one compressed laterally; *face* flat, densely pubescent; *clypeus* subtriangular, gibbous, its base truncated, and apex convexly lobated, or straight and margined; *labrum* lunulate; *mandibles* broad at the base, and obscurely tridentate; *cibarial apparatus* moderate; *tongue* twice the length of the labium, clothed with pubescence to within a brief distance of its apex, and terminating in a small knob; *paraglossæ* about one-fourth the length of the tongue, coadunate at the base, and acuminate; *labial palpi* three-fourths the length of the tongue, broad at the base, and tapering to the extremity of the acute apex of the second joint, which is about one-fifth the length of the first, the two terminal joints very short and articulated laterally just before the end of the second; *labium* one-half the length of the tongue, broadest at its base, and acutely produced in the centre of its inosculation; *maxillæ* as long as the tongue, subhastate and acuminate; *maxillary palpi* two-jointed, short, sometimes equal, and slightly robust, or with the basal joint very robust, and its terminal joint twice as long and linear. THORAX globose, very hirsute, whence its divisions are inconspicuous; *scutellum* lunate; *metathorax* truncated; *wings* with three submarginal cells subequal, or the third the longest, and a fourth slightly commenced, the second receiving the first recurrent nervure near its centre, and the third receiving the second recurrent nervure close to its extremity; *legs* robust, pilose, the four anterior plantæ with a dense, short, setose brush beneath; the *posterior tibiæ* triangular, very smooth, and irregularly concave on their external surface, fringed with long pile along its two external edges, and its extremity tipped with a short pecten of stiff setæ; the *plantæ* elongate and broad, nearly equal, externally shagreened and spinulose, with a longish auriculated process at the external angle of the superior edge, a dense brush of short, stiff hair beneath, and a short pecten of stiff setæ edging its submarginate extremity; the *claw-joint* the longest of the four short subsequent joints, and the *claws* bifid. ABDOMEN ovate or globose, deflected towards its extremity, its base retuse, the last segment triangular, and terminating obtusely.

The MALE differs in always being more intensely coloured; in having the *antennæ* distinctly longer, less distinctly geniculated, the *scape* shorter, the third joint of the *flagellum* almost as short as its basal joint, and the fourth as long as the terminal one, which latter two are the longest of all, and the joints from the fourth to the eleventh severally more or less slightly curved.

NATIVE SPECIES.

1. *lapidarius*, Linnæus, ♂ ♀ ☿. 6–10 lines.
lapidarius, Kirby.

2. *Harrisellus*, Kirby, ♂ ♀ ☿. 6-10 lines. (Plate XVI. fig. 1 ♀.)
3. *subterraneus*, Linnæus. ♂ ♀ ☿. 5-10 lines.
   Soroensis, Kirby?
4. *Latreillellus*, Kirby, ♂ ♀ ☿. 5-8 lines.
   Tunstallana, Kirby.
5. *hortorum*, Linnæus, ♂ ♀ ☿. 5-10 lines.
   hortorum, Kirby.
6. *Soroensis*, Fabricius, ♂ ♀ ☿. 5-8 lines. Plate XV. fig. 4 ♂.)
   Cullumana, Kirby, ♂.
7. *lucorum*, Linnæus, ♂ ♀ ☿. 5-9 lines.
   lucorum, Kirby.
   virginalis, Kirby.
8. *terrestris*, Linnæus, ♂ ♀ ☿. 7-11 lines.
   terrestris, Kirby.
9. *Skrimshiranus*, Kirby, ♂ ♀ ☿. 5-8 lines.
   Jonella, Kirby.
10. *nivalis*, Dahlbom, ♂ ♀. 6-8 lines.
11. *pratorum*, Linnæus, ♂ ♀ ☿. 4-8 lines.
    pratorum, Kirby.
    subinterrupta, Kirby.
    Donovanella, Kirby.
    Burrellana, Kirby.
12. *Derhamellus*, Kirby, ♂ ♀ ☿. 4-8 lines.
    Raiella, Kirby, ♀.
13. *Lapponicus*, Fabricius, ♂ ♀ ☿. 5-9 lines.
    regelationis, Newman.
14. *fragrans*, Pallas, ♂ ♀ ☿. 5-10 lines. (Plate XV. fig. 3 ♀.)
    fragrans, Kirby.
15. *sylvarum*, Linnæus, ♂ ♀ ☿. 6-8 lines. (Plate XVI. fig. 3 ♀.)

sylvarum, Kirby.
16. *Smithianus*, White, ♂ ♀ ☿. 4–10 lines.
17. *senilis*, Fabricius, ♂ ♀ ☿. 6–9 lines.
muscorum, Kirby.
18. *muscorum*, Linnæus, ♂ ♀ ☿. 4–9 lines.
Francillonana, Kirby.
floralis, Kirby.
Sowerbiana, Kirby.
Beckwithella, Kirby.
Curtisella, Kirby.
Forsterella, Kirby.

GENERAL OBSERVATIONS.

These, perhaps the most conspicuous of our native bees, certainly the largest, and probably the most generally known after the domestic bee, have their scientific generic name from βόμβος, an imitative word, made to indicate the sound of the hum of the insects themselves. They have many popular names such as bumble bees, dumbledors, humble bees, and in Scotland they are called foggie bees. They consist of three sexes, males, females, and neuters, which differ considerably in size, the females being very much the largest, and the neuters the smallest. Of course, individually, like all other insects, there is much variation among them in the intensity or diversity of the colouring of their pubescence, from which it is chiefly that they derive their specific distinctions; in the relative sizes of individuals also there are great differences. It is the males, as is usual among the bees, which are the gayest in their attire, and take the widest range of variation, and sometimes so much exceed the typical specific character in their markings as to require experience to identify them, and to place them correctly with their true species, which can only be ascertained with certainty by the examination of the male organs of generation, which differ in the various species, but are undeviating in their specific uniformity. Of this character, which I was the first to discover as being of specific value for critical determination in the separation of the species of very difficult insects, I was enabled to make important use in the genus *Dorylus*, in a monograph on the *Dorylidæ*, an exotic family proximate to the ants, and which was published in Taylor's 'Annals of Natural History' for May, June, and July, 1840. The females and neuters of *Bombus* are less subject to such extensive dissimilarity, and may be usually associated, by their pubescence, in their legitimate groups. Form also frequently lends its aid as subsidiary to their specific identification.

These and *Apis mellifica* are our only social bees, which live in numerous communities under a kind of municipal government which is considerably less perfectly organized in the present genus than in the domestic bee, and thence they are called "villagers," in contradistinction to the citizenship of the hive bee, earned by its comparatively metropolitan institutions, and the centralization of its government, which wholly emanates from the pervading influence of the queen upon the labours, and, indeed, upon the existence of her subjects. But the *Bombi* are under much less social restraint, and admit of several co-regents in the same community, without its being productive of any disturbance of social harmony. In the account of the genus *Apathus*, the last described, we have seen that the *Bombi* are subject to bee-parasites, which in some closely resemble the species they infest, and we have also shown there how these are distributed. The hive bee is not exposed to such intrusion, although, like these, they have many enemies. In the very earliest spring months these *Bombi* are abroad; for as soon as the catkins of the sallow are ripe for impregnation, they are on the wing. But it is now that the large females only are at work, for they have to create their companions before they can be surrounded by them. Their fruition is the result of the previous autumn's amours, at a period too late to form sufficient stores for the numerous brood they will produce, and accordingly, after revelling in a brief honeymoon, they resort, like staid matrons, to a temporary domicile, some cavity just large enough for themselves. In this retirement they pass the cheerless wintry months, requiring perhaps the incubation of time thoroughly to mature their fruit. Whether this be the case or not, as soon as the earth begins to feel the warmth of the sun upon its return from its far southern journey, and to respond to the renewed vitality it gives to vegetation, these bees feel its active influence and come forth. With the progress of the spring and summer most flowers are exposed to their rifling, but they revel upon the elegant flowers of the Horse-chestnut, and their hum is the music of the lime when it is in blossom. According to the species, they select a cavity for their nest, or construct it upon the surface of the ground, this being the case with the CARDER-BEES, which gather moss to construct their residence. In those which inhabit beneath the surface, the selection of an already formed cavity greatly abridges their labour, and their instinct prompts them to choose one sufficiently large for the prospective community, but the nest itself is gradually extended in size suitable to their progressive increase in numbers. All that the parent female does at first is to form a receptacle sufficiently large for her first gatherings of pollen and honey, whereon to deposit her first eggs, and to form a waxen cruse or two to contain the honey requisite for the nest operations of keeping these masses moist enough for the nurture of the larvæ. The material of these pots although called wax is not properly so, but is an agglutination of collected vegetable matter, for it is not plastic to the fingers like wax, and it burns, leaving a carbonaceous residuum very attractive to moisture. The larvæ hatched from the eggs now deposited produce the first neuters, which spin a cocoon wherein they rapidly undergo their transformations. They are, in the first instance, aided to emerge from their silken cot by the parent gnawing off its top, but subsequently this duty is performed, as the family increases, by the neuters then developed. The young bee, on emerging from its cocoon, is not thoroughly hardened in its integument, and its pubescence also acquires by degrees only its proper colouring; all this is not long in being effected, but, until they are thoroughly able to fly forth, they continue to be fed by their elder sisterhood, for the neuters are properly abortive females. Males, and further productive females are produced later in the spring, and are smaller than the normal sizes of those sexes; the autumnal brood, consisting also of males and females, again resume the full size of the complete insect, and it is these females which, after impregnation, hibernate and reappear in the following early spring to be each the parent of a new progeny. The population of these nests varies considerably in the several species: in some, as in that of *Bombus terrestris*, there are more than two hundred, and in that of *B. senilis* there are about a hundred and forty; but it is in those that construct their nests above the ground that the fewest are found. As with the general population, so with the relative proportions of the sexes, the several species vary. Of course all these numbers are approximative only, as under certain conditions they will necessarily differ, nor are the general or relative numbers identical, even in the same species, in the same season, and in the same locality. The proportions are usually somewhat like this, about double the number of neuters to females, and nearly the same number of males as of females. In some of the communities there are even as few as twenty neuters, and these, of course, comprise those species which are most rarely found by collectors. The most pugnacious of all, and the fiercest in their attacks and most painful in their stings, are those which live underground or in cavities formed of accumulations of stones, and it is these which are the least constructive in their habitations, as if their truculent nature rejected the concomitants of incipient civilization; for it is those which build moss-nests, requiring a certain amount of skill, that are the most gentle in their habits. With the increase of numbers in the habitation, the rapidity of the labours progresses, and the accumulations quickly increase; but there is always opportunity for the entire community to find employment, either in enlarging their nests, when they build them, or in securing them from the intrusion of water, or repelling enemies, or feeding the young, and accumulating stores. In collecting pollen they are often covered as if they had rolled themselves in it, and this they brush from their hairy bodies chiefly with their posterior legs; sometimes they return in this disguised condition, and free themselves from it only at home; in other cases they bring it home collected in little masses upon the corbiculum, or basket, of the posterior shanks. They may be often caught thus laden, and I once captured a large female of *B. terrestris*, with the shanks and plantæ of both intermediate and posterior legs covered with masses of thick clay, required doubtless at home for some domestic repairs. The instinct of these bees teaches them that where the tube of the flower is too narrow for the introduction of their body, and too long for even their long proboscis to reach the nectarium at the bottom, they may get at the honey by piercing a hole near that organ, which they know where to find, and thus they readily get at the treasure that they seek, lapping it through the aperture and carrying it off. If, in their collecting-excursions, they are intercepted by heavy rains, or loiter far away too long until the twilight closes, they will pass the night away from home, and return laden with their gatherings as soon as the warmth of the sun reanimates them to activity; thus they will often sleep in flowers, and a nest therefore taken at night is not always a sure indication in those found within it, of its complete population. In their amours, the autumnal females evince considerable coquetry to attract their partners: they place themselves upon some branch in the most fervid sunshine, and here they practise their cajoleries in the vibrations of their wings, and allure them by their attractive postures, The males are simultaneously abroad, and soon perceive them. The seduction is complete, and they pounce down upon them with impetuosity, but their brief indulgence terminates in death, for with his abating vigour the female repulses him, and he falls to the ground never to take wing again. Amongst their insect enemies the *Dipterous* genera, *Volucella* and *Conops*, are very destructive to their larvæ,—the first of these genera in its colouring greatly resembling the species upon which it preys. Foxes, weasels, field-mice, all prey upon them, and, like schoolboys, often destroy the bee for the sake of its honey-bag, an instance of which I have before recorded as illustrative of their endurance of the loss of a considerable portion of the body without its being fatal.

The most interesting part of their history is perhaps that upon which I have not yet enlarged, namely, the structure of their nests. This is particularly the case with the carder-bees, which felt and plait the filaments of moss to form its whole enclosure. Such species select a spot close to an abundant supply of the material; this they bite off and form pellets of. To these nests a moderately long arched passage is formed of the same material, of sufficient size to permit the free passage of the bees to and fro. This necessarily is shorter at first and leads to a smaller receptacle when the parent bee works alone. But as her offspring of workers increases, the passage is lengthened and the nest enlarged. To construct it, when in full activity, the bees form a chain, one behind the other, extending from the growing material to the entrance of their passage to the nest, all their heads being turned towards the moss and their backs to the nest. The first bites off the raw material, rolls it and twists it, and passes it to the second, by whom and the succeeding ones it undergoes further manipulation, and where the chain terminates at the commencement of the passage another bee receives it and conveys it along this into the interior, and then applies it itself or passes it to others thus employed where it is required. A vaulted covering and sides is thus formed or extended within the cavity by the plaiting or wreathing together of these sprigs of moss, and the inside of which is further strengthened by being plastered with a coating of the pseudo-wax, which, however, smells much like true wax, and with which the lower loose filaments of the moss are intermingled, that one cannot be separated from the other without tearing the whole to pieces. Thus ingeniously do these insects enclose their home. These nests are not always on the surface, but often cavities of the necessary size are thus lined, and then they are doubly secure. Within these nests, with the increase of the population the number of the cocoons of course increases, as they are never used twice over, excepting that when they are conveniently situated for the purpose they are converted into honey pots. Thus sometimes several layers are formed of these irregularly-placed cocoons, of which the longest diameter is, however, always perpendicular to the horizon. In this way *B. muscorum, senilis, fragrans,* and others build. Some use a naked cavity, and merely secure it in its crevices from the filtering intrusion of rain or other water, the closing patches being formed of the usual waxy material. This is the practice of *B. terrestris,* which associates the largest communities of all; and *B. lapidarius* seeks cavities among stones or in the earth, and forms a nest of a regular oval, but merely clothes the sides, which is done by bits of moss and grass carried carefully home. The domestic arrangements within are much the same in all, the prolific females and the neuters being the labourers, which perform all the duties of building, the collecting and caring for the young, the function of the males being limited to the perpetuation of the species.

Subsection 2. WITHOUT SPURS TO THE POSTERIOR TIBIÆ.

‡‡ *Permanently social.*

Genus 27. APIS, *Linnæus.*

(Plate XVI. fig. 4 ♂ ♀ ♀.)

APIS ** *e* 1, Kirby.

Gen. Char.:—THE NEUTER.—BODY nearly cylindrical and subpubescent. HEAD transverse, about as wide as the thorax; *vertex* and *face* deeply longitudinally channelled in the centre, the latter to the apex of a small triangular elevated space between the insertion of the antennæ, and extending to the base of the clypeus, the sides of the face flat; the *ocelli* rather large, seated far back upon the vertex in a triangle, the anterior one in the depth of the longitudinal channel, the two lateral ones placed further back towards the occiput in a transverse indentation crossing the longitudinal one; *compound eyes* very pubescent; the *hexagonal facets* very minute; *antennæ* short, filiform, geniculated; the *scape* nearly half the length of the flagellum and subfusiform, the basal joint of the flagellum globose, the second subclavate and subequal with the remainder, very slightly lengthening to the apical joint, which is compressed and as short as the second; *clypeus* quadrate, convex; *labrum* transverse, linear, slightly waved in front; *mandibles* broad at the apex, edentate, obliquely truncated and concavo-convex; *cibarial apparatus* shortish; *tongue* nearly twice the length of the labium, linear, pubescent, and terminating in a small knob; *paraglossæ* obsolete, coadunate with the base of the tongue; *labial palpi* not quite so long as the tongue, the first joint four times as long as the remainder, and tapering from the base to the apex of the second joint, which is about one-fourth the length of the preceding, and has the two very short terminal joints articulated just before its acute apex; *maxillæ* broad, hastate; *labium* half the length of the tongue, its inosculation straightly transverse, not so long as the tongue and acuminate; the *maxillary palpi* extremely short, the basal one the shortest. THORAX subglobose; *prothorax* inconspicuous; *scutellum* lunulate and impending over the post-scutellum, which is transverse and linear; *metathorax* truncated; *wings* with a long marginal cell extending nearly to the end of the wing, and obtuse at its extremity, three submarginal cells which terminate at less than half the length of the marginal, the second the largest and receiving the first recurrent nervure towards its commencement, the third oblique and narrow and receiving the second recurrent nervure just beyond its centre; *legs* slender, subpilose; the anterior and intermediate *tibiæ* with a spur, their *plantæ* with a dense short close brush all round, the *posterior tibiæ* triangular, glabrous within, externally smooth, shining, and irregularly concave, the edges fringed longitudinally with long hair curving inwards, and forming the sides of the corbiculum, or basket, which conveys the *matériel* of the nest, the apex transverse and pectinated with short rigid setæ, but wholly without spurs; the *plantæ* oblong, not quite so long as the tibiæ, the sides nearly parallel, the upper edge fringed with long loose hair, subglabrous externally, but furnished internally with ten transverse, parallel rows of short stiff golden hair, with an auricle at the outer angle, forming collectively a dense brush, and its oblique apex pectinated with short stiff setæ, the remainder of the tarsal joints short, the fourth the shortest, and the claw joint the longest; the *claws* short, robust, and bifid. ABDOMEN retuse at the base, subcylindrical, convex above, and terminating conically, the first segment very short, the second the longest, the ventral segments ridged longitudinally in the centre.

The FEMALE, or QUEEN differs in the head not being quite so wide as the thorax, in having the *cibarial apparatus* very much shorter; the *mandibles* distinctly bidentate, the inner edge of the inner tooth stretching obliquely to the acute inner extremity of the broad apex of the organ; the *labial palpi* as long as the tongue, with all the joints conterminous, the basal one slightly acuminate, the second linear, the two terminal ones more slender and shorter, the pubescence of the eyes very much longer than in the neuter; the *legs* more robust and less pilose; the *posterior tibiæ* convex externally, without the lateral fringes of hair, and their plantæ merely oblong, without the external basal auricle. The ABDOMEN is also considerably relatively longer; and has not the central ventral ridge.

The MALE or DRONE differs from both in being considerably more robust and more completely cylindrical, and very much more densely pubescent; the *compound eyes* contiguous at the summit, occupying the whole of the vertex, and nearly all the lateral portions of the face, extending below to the articulation of the mandibles, their pubescence much shorter but denser than in the other sex; the *ocelli* large, and seated at the top of the central portion of the face in a close triangle, a little above the insertion of the antennæ, and in front of the conjunction of the compound eyes, the lateral ones of the triangle being closely contiguous to the upper inner edge of those eyes; the *antennæ* are more robust and rather longer; the *cibarial apparatus* very short; the *labial palpi* about three-fourths the length of the tongue, and the joints conterminous, the *tongue* robust; the *thorax* is nearly quadrate; the *legs* are nearly naked, the four anterior very slender; the *posterior tibiæ* slightly curved, convex externally; the *posterior plantæ* more robust, and more convex externally than their tibiæ, they are regularly oblong, and without the basal auricle, the rest of the joints of the tarsi are very short. The ABDOMEN robust, and obtuse at its extremity, but its seventh segment is concealed beneath; the *ventral segments* concave longitudinally.

NATIVE SPECIES.

1. *mellifica*, Linnæus. (Plate XVI. fig. 4 ♂ ♀ ♀̣.)
mellifica, Kirby.

GENERAL OBSERVATIONS.

The name of this genus, *Apis*, adopted by Linnæus as the classical generic name of the bee, although with him it comprised the whole modern family of these insects, but which, as now restricted, in accordance with its limitation exclusively to the congeners of his adopted type, is the ancient Latin vernacular name of the honey bee, and to which it has been ever since uniformly attached. This name, as shown by its derivative meaning, was originally imposed with direct reference to the insect's constructive habits, as was the case with the names given to it in the more primitive languages before referred to, and which is also the origin of its Teutonic and Scandinavian appellations—*Biene, Bie,* and *Bi,* whence our own common name for it is obtained through the Saxon *Beo,* and we have beside Bye or bee, signifying *a dwelling.* From this circumstance it would seem that a very early and universal discernment existed

Y

of its ingenuity and skill, its significant name being everywhere analogous.

The habits and economy of these industrious little creatures have been a source of greater wonder and admiration the more closely and accurately they have been observed. They have attracted the thoughtful speculation of minds of the largest compass throughout all ages, which, reasoning upon the *modus operandi* of these insects, have endeavoured to define, and determine the differences between instinct and reason, with their precise limitations. But baffled in their attempt to settle whether these be affinities or analogies, it should rather have persuaded them to adopt the motto of Montaigne, and exclaim, *Que sais-je?* Into these metaphysical discussions it is not necessary to enter, and I confine myself to the natural history of the insect.

Although the description of the three sexes which comprise the population of the hive are technically given above with scientific precision, it will be as well, perhaps, to recapitulate them briefly, with their distinctive attributes, in a more popular form.

They consist of a queen, or productive female, whose function is thought to be exclusively to lay eggs, but who may perhaps have some hitherto undiscovered control over the executive of the hive, to be implied by the confusion invariably following her death or her removal from the community, and which becomes totally destructive to its organic constituency unless stayed by another monarch being improvised, or by one extraneously supplied; one monarch alone rules without a coadjutor, and without any equal being tolerated, for the presence of a second queen, or the immature larva of one, even of her own progeny, maddens her to murderous aggression, or to the impulse of emigration accompanied with a host of adherents. She never leaves the hive when once her duties have fully commenced, for by distinction of structure she is rendered incompetent to execute any of the labours that devolve upon the workers; her tongue is formed only to lap nutriment; she has no cysts for the secretion of wax, she is without the honey-bag for conveying that liquid home, and her posterior shanks are convex externally, and thus deficient in the concave basket for carrying home the stores of pollen or propolis, whilst their plantæ are without the little earlet at the top externally, or the close dense brush arranged in rows within, which aid these workers in their many manipulations. Her wings are too short to convey her ponderous body through the air, and her sting becomes stronger by being curved. Thus she is exonerated from labour by the incapacity of her structure to execute it, although her duties are quite as incessant and as arduous, being indispensable to the perpetuation of the species.

Her consort, the DRONE, is the male of the hive, and although the queen is monandrous or single-spoused, and although the hive during the season rarely throws off more than three swarms, usually restricted to the accompaniment of a single queen, and thus but three males are absolutely required, nature is so provident of the great design of perpetuation, that to provide against the possibility of its frustration, the hive usually produces about a thousand drones. A peculiarity in the structure of the drone which facilitates his discovery of the virgin queen when she issues from the hive on the bridal excursion, which she makes preliminary to her heading a swarm of emigrants, or assuming monarchy at home, consists in the vertical enlargement of his compound eyes, which meet over the brow, and in the posterior expansion of the inferior wings, which take a broad backward sweep, giving the insect larger powers of flight, but perhaps required as much by its own bulkiness and weight as for the purpose of ascending above his bride in the upper regions of the air; but that its weight cannot be the sole reason is testified by the analogous structure in the male of the genus *Astata*, one of the fossorial *Hymenoptera*, where a similar expansion of the inferior wing is concomitant with a similar development of the compound eyes, yet in which the abdomen is very small, and this power is therefore evidently given to these merely to increase the velocity or the duration of their flight. The rest of the structure of these drones disables them, like all other male bees, for any labour; and as they must be sustained as long as they may be of service, the possibility of which terminates with the last issue of a swarm from the hive, a period appreciated by the instinct of the workers, they are then driven forth, but it is in dispute whether the workers destroy them, or whether their destruction is effected by exposure and hunger, or by the natural limitation of their lives, for although their tongues are formed upon the same type as that of the worker, it is considerably less developed, and appears to be adapted only to obtain nutriment from the honey already collected in the cells, as they seem even deficient in the instinct to gather it for themselves from flowers, never being observed to visit them.

The last inhabitant of the hive is the WORKER, or abortive female, whose labour has several phases. A difference of size amongst them has been supposed to have been noticed by observers as varying with their occupation and duties, but as they are all constructed in the same manner, with precisely the same organs, which are of the same form and in the same situation, this must be a mere imaginative surmise. Their similarity of structure permits them, collectively, to apply themselves to the same occupations which the needs of the community may at any moment demand. Taking them separately with their distinctive occupations at any given time, without implying by it a permanent separation of classes, we find them to consist of wax secreters, builders or cell-sculpturers, honey collectors, pollen collectors, propolis collectors, nurses of the young, ventilators, undertakers to carry off the dead, who are perhaps also the scavengers which cleanse away any occasional dirt, sentinels to guard the hive outside and inside, and attendants upon the queen, or as the "'Times' Bee Master" very aptly designates them "ladies in waiting," and at all times many slumberers are reposing from their toils. That all these duties are transferable, and consequently are transferred indifferently from one to the other, is implied by their general capacity for fulfilling them resulting from this identity of structure, which will be understood as not at all infringed by the separate capacities I unfold as devolving from their temporarily limited functions, all being simultaneously in action, but distributed amongst the several individuals.

The first important occupation of the worker is the secretion of wax for the structure of the cells, and, to effect this, honey must be collected, for it is solely from the digestion of honey that the wax is produced. This in due course passes from the first stomach or honey-pouch wherein it is collected, thence to the second stomach, and then on to the cysts or little bags which run along on each side from the second to the fifth ventral segments, and correspond and communicate with eight trapezoidal depressions placed externally upon the plates of the ventral segments—four on each side, through the concavity of which the secreted wax exudes in a liquid, transparent, hot state, forming a thin scale within each, which the air hardens into a white substance, as the pulp of paper is hardened upon the form into which it is introduced, or like salt crystallizing into flakes from sea-water in shallow salines. This, however, is not yet wax, although its essential constituent, but to become so these scales are removed by the scopulæ of the posterior plantæ and their auricle, to the intermediate feet and by these transferred to the anterior pair, which pass them to the mandibles, where they are masticated and mixed with a saliva issuing from the mouth, and thus intermingled they consolidate into a white opaque mass, which issues from the mouth like a thin strip of riband, and constitutes true wax, plastic to their manipulation. To form this secretion, the bees having collected the honey themselves in the first instance, or having consumed sufficient before leaving the hive with the swarm, but which they subsequently obtain from the supplies stored in the present hive, hang themselves in festoons in all directions about its cavity, each festoon being formed by two parallel chains of bees clinging together; the top bee on each side hangs by its anterior claws to the top of the hive, and the next in succession grasps with its fore claws the hind claws of that and so on, until the depth of the festoon they find to be sufficient, when the bottom bees of each chain swing themselves together, and cling to each other in the same manner by their hind claws only. These festoons are speedily suspended, and with a fresh swarm are in immediate active operation. The secretion requires about twenty-four hours to complete, and as this is accomplished the festoons break up, and these secreters convey it to where the sculpturer bees or builders are moulding the cells, to whom it is successively supplied by the secreters themselves as wanted, for none is stored, although the wax of old or dilapidated parts of the hive, or of the vacated cells of the new-born queens are reconverted to use. These builders are very rapid in their construction of the hexagonal cells, which, as they are progressively completed, are stored with honey, this being during the time assiduously gathered by the honey collectors, and these cells are interspersed occasionally with those wherein pollen or propolis is stored, each of which, as the bees collecting them successively return, is cast into the selected cell by the bee collecting it, who returns at once to the same employment, whilst the store thus deposited is immediately compactly pressed in and warehoused by other bees who fulfil that duty, or who cover it in when the cells are filled, with a waxen covercle formed of concentric circles; or, in the case of the honey-cells, to keep the thickened operculum deposited upon it in due position and repair, after the retiring of the bee which brought home the fresh store of honey, and which had displaced it to regurgitate her addition into the cell. This operculum or cover is of a thicker consistency than the honey itself, and prevents its oozing from the cells, which would often take place from their uniformly horizontal position, were it not for the sagacity which prompts them to introduce this preventive, and which is not removed until the cell is filled; it is then covered hermetically with its waxen top.

A sufficient number of cells being ready, and sufficient stores of honey, pollen, and propolis for the progressive labours of the hive, and a great number of empty cells all finished for the use of the queen, she begins to lay her eggs. As these are hatched the duty of the nursing-bees commences, which is to feed the young, who crave for food like young birds, and are as diligently supplied by these nurses with a material called bee-bread, which consists of masticated pollen, the pollen being exclusively stored and used for the purpose. This is mixed with some secretion from the mouth, which converts it into a sort of frothy jelly. These bees are never negligent of their duties, and with their feeding the larvæ rapidly grow.

To keep up a necessary supply of air in the hive, and to prevent suffocation from heat, a certain number of the community are employed in fanning the passages between the cakes of comb and the whole interior of the hive, by the vibration of their wings, which thoroughly ventilates it, and the accumulation of deleterious air is prevented; some, for this purpose, being posted at the aperture to the hive, where, this vibration causing a temporary vacuum, the external air rushes in, and the chain of succession of bees within becoming thus vibrating air-valves completes the ventilating arrangement. While all these operations are progressing, a certain number are acting as a militia of citizens, who have substitutes only in the succession and change of duties. These act as sentinels, who guard the entrance and patrol the interior and courageously intercept all inimical intrusion, for the bees have many enemies, but who are merely so to benefit themselves, and are not parasites of the nature of the bee parasites of the solitary kinds; and where they cannot individually avert it, they obtain collateral aid from others of their staff. The next class is the attendants upon the queen: these vary in number from twelve to twenty; they invariably accompany her wherever she proceeds throughout the hive, for the purpose of laying her eggs; and whether their custom gave rise to the etiquette which attends human royalty, that a subject may never turn the back upon the sovereign, these attendant bees surround her with the head always turned towards her, and seem to caress her with their antennæ and pay her every kind of deferential homage, those in front moving backwards as she advances, and those on each side, laterally, so that they ever face her; and as they tire others succeed them in their duties. Another set fulfil the office of keeping the hive thoroughly clean, for the transit of such large numbers will inevitably collect occasional dirt, as will the drift of the wind at the entrance of the hive and the action of the ventilators themselves. Their duty it is also to remove any extraneous organic body that has forcibly entered and which may have succumbed to the vindictiveness of the bees. Where they are not strong enough, even collectively, to effect the removal, as in the case of a mouse or anything else as large or larger, they then call to their aid the wax workers and the repairers; these enclose the obnoxious body, which they have the judgment to know will become dangerous from putrefaction, to aid in its prevention, by a cerement of wax or propolis, which prevents any offensive exhalation, and thus secures the wholesomeness of the hive.

Here is completed, with the enumeration of those which successively repose from their toil, the several labours of the community which inhabits the hive.

The structure of the workers, which enables them to carry on all these operations with the requisite facility, is very different from that of the two sexes we have just described. As before said, they are abortive females, but, as I shall have occasion to explain lower down, capable of having this special incapacity removed, if the necessary process requisite to be adopted for the purpose be applied within three days of their being hatched into the larva state. The acquisition of the faculty of fertility entails, however, the loss of all power of pursuing any of the other occupations of the hive practised exclusively by the workers in general. The nurture that gives it them converts them into queens, and moulds them to the structure of this sex described above. As a remarkable and rare exception, some one or other of these workers will occasionally have power of laying a few eggs, but which are always those of drones. The other peculiarities of their structure are its adaptation to the secretion of wax above described; and their power of throwing up the honey they have collected in the first stomach or honey-bag, before it passes on by digestion, somewhat in the way the ruminant quadrupeds bring up the cud, of course by muscular action, without the convulsion of vomiting. Their next distinction is that their mandibles are edentate and more like spoons, and are often so used, or as the plastering-trowel of masons is for smoothing surfaces. Their legs remarkably differ from those of the other sexes, all of their limbs being somewhat adapted to the collection and conveyance of pollen and its manipulation, as well as that of propolis; but it is the posterior shanks which are specially constructed for the conveyance of these materials, by being framed externally like a little basket; being hollowed longitudinally and their lateral edges fringed with recurved hair, which retains whatever may be placed within the smooth and hollow surface, and the apical extreme edge has a pecten or comb of short stiff bristles. The first joint of the posterior feet have also their distinctive form, adapted to special branches of their economy. These are oblong, wider than the shank, and about two-thirds its length, and consequently powerful limbs; at the outer angle of the edge, nearest the shank, is a little projection called the auricle or earlet, the inner surface is clothed with ten parallel transverse rows of close dense hair, and its apical edge has along its whole width a pecten similar to that of the apex of the shank. This shank being without spurs, which only the domestic bee is deficient in, gives the pecten a freedom of action it would not otherwise have, and enables it to be used together with the earlet opposite to it on the foot, as an instrument for laying hold of the thin flakes of wax upon the venter, and to bring them forward to the intermediate legs to be passed on to the mouth, and there to be converted into wax. The pecten of the foot and also its brush aid in their removal in case of need, and help as well both in the manipulation and the storing the materials collected. Thus, this whole structure, exclusively possessed by the worker, is pre-eminently designed for the manifold operations of the hive; and the bee itself and its works are but one closely linked chain of wonderful contrivances.

The entire economy of the hive seems to emanate exclusively from the two most prominent attributes of instinct, that of self-preservation, and that other more important axis of the vast wheel of creation, the secured perpetuation of the kind by the conservative στοργη, or absorbing love of the offspring. The latter is more eminently developed in the social bees than in any other group of the family of these insects. In the solitary bees it presents itself as a blind impulse, unconscious of its object; for did we admit the consciousness of the purpose of their labours, we should evidently endow them with reason. How could they know, without reflection, that the food they store in the receptacle they form for the egg they will deposit, and which receptacle is exactly adapted to the size that the larva which will be hatched from it will take, is to nurture a creature they will never see, and whose wonderful transformations they will not therefore witness? In the hive bee the maternal instinct exhibits itself as an energy diffused though a multitude of individuals, but these witness the results of their solicitude, and exclusively promote its successful issue; and in these also the instinct of self-preservation is a diffused impulse, which likewise includes the preservation of the society.

As male and female conjunctively make up the species, thus do the queen-bee and the neuters collectively make up one sex,—the mother,—for the functions performed by the female alone in the case of the solitary kinds of bees are, in the genus *Apis*, separately executed. The cares and labours of maternity devolve upon these neuters, while the queen-bee's maternal function is limited to merely laying the eggs with which she is replete, with the instinctive power of selecting for them their proper depository,—each of which is adapted in size to that of the sex which will be produced. Her maternal instinct stops abruptly here, without the development of an afterthought or care for their future thriving. The instinct of the neuters, like the anticipative promptings of the human mother, to prepare the clothing and other necessaries for her expected infant, has forecast the queen's needs in its intermittent urgency, by progressively constructing cells fitted severally in size for the growth and nurture of neuters, the first developed; of drones, the next produced; and lastly, of queens, which soon afterwards appear; she instinctively knowing the proper time and the suitable use of them, having the faculty of distinguishing them with a view to the deposit of the particular kind of eggs of which she is for the moment parturient.

The drones, or male bees, appear to receive life for one substantial purpose only, which is soon accomplished, but during the short space of time its successive performance requires, it is incidentally accompanied with assistance to the general community whilst they remain permitted occupants of the hive, by aiding in heating and ventilating it,—a labour repaid by the food, which they obtain from the stores kept open for daily consumption. Although uncontributive to the acquisition of the riches of the hive, yet are they indispensable to the perpetuation of the species, and their murder as supposed by some apiarians, or their expulsion as thought by others, in either case equally terminating in their destruction, seems an unworthy return for the important service performed, although this is restricted to the number of individuals required by the equal number of queens that may be produced. To this number their production might be limited, but for the chance of either or all of these queens failing by some casualty to obtain a prince consort. To baffle the possibility of this mischance, a very superfluous number of these drones is hatched, as above stated, which are on the alert, when each queen successively issues forth upon her bridal morn, to catch her favouring glances, and be the accepted groom. That they are not further conducive to the well-being of the hive is the fault of their structure and of their instinct, which are correlative, they being as little fitted either in their tongue or their legs for the uses of the hive as the queen herself. The physiology of their intercourse is a mystery of mysteries, and would seem to partake of the principle, modified, of that developed in the aphides, where the vital power passes on through successive generations by the efficiency of the energy of one ancestral intercourse. In the hive-bee this is not the case, but in these the one espousal fertilizes eggs to the number of often a hundred thousand, yet undeveloped and even indiscernible by the aid of the microscope in the ovaries of the queen, and which become bees progressively in the course of a couple of years, the supposed duration of her existence, during the whole of which time she is laying. The accepted male is destroyed by the effects of the amour, and when all the queens which are to be the heads of independent communities are successively fertilized, and have led forth their colonies, the remaining drones issue compulsively from the hive and are lost in the wideness of nature, and die by the natural limitation of their existence, or become the prey of their numerous enemies.

The neuters or workers are, as it were, emanations of the queen, or the organs whereby her several functions as a mother are performed, considering the species as restricted to two sexes, and thus they comprise with her, collectively, one organic whole. That this is a consistent view of their condition is further proved by the circumstance that from their larvæ, upon the failure of a queen, a new queen is produced upon one being supplied with a certain nutriment that developes the capacity that would remain inert and abortive, were it not thus promoted from its primary state. It may be questioned whether the eggs deposited by the queen in the royal cells are other than neuter eggs, their subsequent nature being changed by the different quality of the sustenance they are fed with when hatched, as is the case in the above noticed defection of a queen. This then would limit the queen's eggs to the eggs of neuters and of drones, thus further corroborating the idea of the existence of but two sexes.

I have stated above the supposition that the queen's office may be restricted to the laying of eggs, but it must be inferred that it has a wider compass, and possibly comprises some administrative function in the regulation of the hive, from the circumstance that with her loss the entire community loses its self-possession and self-control. Labour then ceases and the hive becomes the scene of turmoil and confusion, and unless the loss be repaired in the way named above, which their instinct teaches them to adopt, if any eggs have been already deposited, or if supplied by the surreptitious introduction of another queen which they immediately raise to their superintendency, paying her the same deference they had done to their lost monarch, or would do to a legitimately native birth, it disperses and destroys the community. Such a loss in its natural course must necessarily, to be effectively repaired, take place in the interval after the laying of the drones' eggs, and before those of the queens are deposited, for otherwise she would remain unimpregnated. Having thus shown reasons for supposing that the hive actually contains but two sexes, and having also shown that the first phase exhibited of this distributed maternal instinct by which the neuters form conjunctively with the queen a many-headed and many-hearted mother, is their preparation of the cells for all the purposes required,—the next and most important, and the one perhaps which elevates them vastly higher in the scale of social intelligence and affection, is the absolute development in them only of maternal solicitude for the well-being of the offspring. This certainly proves the existence of the diffused maternity urged, for they feed the hatched young as the bird does its callow, from hour to hour, and which, when full grown, they enclose in its formative cell, to undergo its changes and become one amongst themselves. It is not absolutely determined whether the functions performed within the hive are restricted to distinct sets of the workers, but it may be presumed that the duties are transferable, for the most plausible supposition is, that all the offices are interchangeably performed by the entire population, possibly merely limited to daily alternation of individuals taled off each morning for the day's duties. That an administrative regulation must exist under some executive authority, emanating doubtless from the centralization of all in the queen, and communicated to the rest by her relays of attendants, may be conclusively inferred, otherwise all might similarly employ themselves from day to day, and thus overwhelm with one work the multiplicity of labours required for the well-being of the hive. For whilst some are secreting the wax from the honey they have consumed, others are moulding it into shape, others are harvesting the bee-bread to feed the voracious larvæ, others are gleaning the propolis for the security of the

domicile, others are collecting honey to store as needful
supplies, others are either ventilating or heating the interior, others act as sentinels and guard the approaches
or patrol the passages within, and will die in that defence
like genuine patriots, and others are in attendance upon
the queen in her progresses through her dominions, and
who may individually act as *aides-de-camp* to convey her
commands to the rest. All these are not fanciful em-
bellishments of the narrative, but substantial and well-
authenticated facts, supported by the repetition on many
sides of careful observations, but perplexing to human
intelligence, for not the least wonder of this conventicle
of wonders—the hive—is that it confounds the astute
reason of man to comprehend it in all its significancies.

The first necessity of a new colony is the selection of
a locality for habitation, which is usually effected by preliminary trustworthy intelligencers determining upon a
site suitable from its concurrent conveniences. A sufficient supply of sustenance must be conveyed by the
emigrants to accompany the preparatory construction of
the settlement, until land can be cleared, grain grown,
etc., and a year at least will pass, even under the most
favourable circumstances of the exertion of the greatest
industry, concurrently with the most propitious succession of the seasons, before it can become self-sustaining.
But when once the wheel is fairly on the move, round it
spins without interruption or relaxation. The colony
thrives, increasing rapidly in its population; and where
all have put the shoulder to the wheel it climbs the steep
and rugged hill of prosperity, whilst those who are carried onward by its evolutions, from each of the many
successive terraces of this noble height, survey a broad,
cheerful, and fertile landscape, extending itself with their

z

elevation, spread out to a distant horizon, which many of the more venturous spirits amongst them, urged by the teeming increase of their compatriots, have already traversed, and who themselves are now rejoicing in the establishment of offshoots, which speedily rival, in successful fruitfulness, the wide-branched productiveness of the parent stock.

This is strictly the history of the hive, and the parallelism is complete, even to the conveyance with them of the preliminary needful stores. Before a swarm issues from the hive, some fly forth to select a dwelling-place, and return, it is presumed, to make their report.

The population of the hive becoming so dense that there is no longer room for the free and unrestrained circulation of the ordinary processes of the community, and so hot from the inconvenient accumulation of such numbers,—for they extend sometimes to as many as fifty thousand,—instinct prompts a portion of the community to migrate. This disposition is further promoted by the progressive, or completed development of some of the young queens. The inveterate and internecine animosity of these—anticipated rivalry, suggesting, it is surmised, the murderous desire, but being prevented from its indulgence by the defensive guardianship of several of the workers—urges the old queen to abandon at this conjuncture her royal metropolis. The inclination to do so, it would appear, is already foreseen by a very large body of her subjects, for if her departure be delayed by her successor's protracted incapacity for undertaking the sovereign rule, the intending emigrants, having already abandoned all the labours of their old domicile preparatory to their issuing forth, will cluster in groups about the bee board until she is ready to emerge.

This condition will sometimes last a day or two, and thence of course all is confusion both within and without the hive, for her subjects have suspended their labours and she has suspended her egg-laying, and roams wildly about within, striving, whenever she approaches a royal cell, or a fully developed young queen, to attack the latter, and destroy her by stinging her to death, or, to tear the former to pieces to get at the imago within, which indicates its apprehension by a shrill piping sound. But she is forcibly dragged back from this apicidal purpose by the working bees which surround each, and who now intermit their usual deference to prevent this destruction, and bite her and drag her back. The future queen of the abdicated throne having, during this turmoil, returned from her wedding tour, and being still protected from slaughterous aggression, the old queen indignantly issues forth. This exodus takes place usually on a brilliant and warm day, between twelve and three,—accordingly during the hottest hours. This is the first swarm of the year, and if the season be very genial it will take place in May. In this migration she is accompanied by all her most faithful lieges, which comprise, to the honour of beehood, by very much the largest majority of the inhabitants, to the number usually, in a well-stocked hive, of several thousands,—say from ten to twenty, depending on the population of the hive.

Having thus issued forth in a body, they shortly alight upon and about the branch of some adjacent tree, clustering, in as close proximity as they can, to their royal leader. In a natural state, when duly organized to proceed, they would thence start for the domicile that had previously been selected by the emissaries above noted;

but, as their natural habits are not at all perverted by their subjugation to man, we will pursue their history under his dominion. This will be the more convenient, for in the comfortable hive to which they have been transferred by his agency, we shall have every opportunity of exactly watching their manœuvres by the facilities yielded in its being glazed for the purpose. We shall thus be enabled to see and follow the wonderful economy of the hive and its many mysteries, which it would not have been possible to accomplish in an abode of their own choice,—some cavity presented by Nature herself, the hollow of a tree, or an excavated rock. They are, therefore, now housed, and after the survey of the capacity of their abode, which is a short affair, with all the prompt energy peculiar to them they at once commence their labours. The queen is already matured, and ready to lay eggs. In a natural abode the gathering of propolis would perhaps be a first necessity to make their home water-and-wind-tight, for they abhor the inconveniences of the intrusion of wet or cold. It is with this material that they make repairs, fill crevices, and strengthen the suspension of their combs, which are hung vertically; and they apply it also to other purposes, which we shall see hereafter. This material is of a resinous nature, it has a balsamic odour, and is of a reddish-brown or darker colour, and is supposed to be collected from fir or pine trees, or from the envelopes of the buds of many plants, or their resinous exudations, especially that of the blossoms of the hollyhock. It is exceedingly clammy, and they have been observed ten minutes moulding it into the lenticular pellets in which they carry it home in the corbicula, or little basket, of the posterior tibiæ. They gather it like pollen with the fore feet, and pass it to the intermediate ones, whence it is taken by the posterior plantæ, kneaded into shape, and deposited upon the hind shanks. It dries so rapidly that often, upon arriving home, the bees which store it have much difficulty in tearing it from the legs of these collectors. The hottest days only are propitious to its gathering, for all moisture is injurious to it, and the hottest period of the day, also, is alone occupied in its collection. It is said that they have been known to fly as many as from three to five miles for it, from the circumstance that suitable plants were not to be found within a lesser radius; but this may be a mistake, for their ordinary excursions are not supposed to range wider than a single mile or something more, and bees may be able to find it where we may suppose it not to occur. In the abode with which we have provided them it is not so urgent a necessity, this being already wind-and-water-tight, although in the progress of their labours they find it indispensable, and use it to fasten the crevices that intervene between the bottom of the hive and the bee board, and, as before noticed, to strengthen the support of the cakes of comb which hang from the roof. The name it still retains is that which was applied to it by the ancients, and signifies *before the city*, as indicative of its use in strengthening the outworks.

Conjoined herewith is the imperative need for the construction of cells for every purpose of the hive, namely, for the storing of the propolis, and that of the pollen, as also the collected honey, as well as for the reception of the young brood, for the mature queen is waiting impatiently to deposit her eggs. Simultaneously, therefore, is the wax being secreted and elaborated by the processes previously noticed. The community is already large, and all are at once in active operation, but four-and-twenty hours must elapse before the cells can be commenced, for it takes that time to secrete the first batch of wax. Festoons, as before described, of these wax secreters are hanging in every direction within the cavity of the hive, and as soon as the process is completed by the first festoon, this dissolves itself by the several bees unlinking their feet, and a leading bee proceeds to the top of the centre of the hive, where she makes herself room from the lateral pressure of other bees, by turning herself sharply about and agitating her wings, and there she collects the scales from the surface of her ventral segments, manipulates them as before noticed, and thus converts them into wax. The rest follow her, and she collects it from them into a little oblong mass of about half an inch; whilst other bees from other festoons are continually arriving to deposit their produce; and as soon as the mass is sufficiently large, which is speedily the case, a sculpturer bee succeeds, and the first cell is laterally commenced. On the opposite side to where this is being framed, two other bees are at work, moulding the bottoms of two cells in apposition to the basis of the first one. The wax keeps constantly increasing by fresh deposits, and the rudiments of more cells are as rapidly formed. These all emanate laterally, in a horizontal direction or with a very slight incline towards their base. They gradually form the vertical cake of comb, for the bottom of one entire range of cells suffices for both sides and inevitably they are so adjusted that the bottoms of those on either side are each covered by one-third of the bottoms of each cell on the opposite side, and so conversely, receiving and communicating strength by three thus supporting one. Here comes the great wonder of the hive; here in this fragile structure abides a mystery that has perplexed man's keenest sagacity. Is it accident or is it intelligence that instructs the bee, or is it the impulse of the instinct implanted by that Supreme Intelligence which gives man his reason and moulds all things to their most fitting use?

Ray's view is precisely this; he says:—"The bee, a creature of the lowest forms of animals, so that no man can suspect it to have any considerable measure of understanding, or to have knowledge of, much less to aim at, any end, yet makes her combs and cells with that geometrical accuracy, that she must needs be acted by an instinct implanted in her by the wise Author of Nature." To support this idea of the geometrical skill of the bee, he cites "the famous mathematician Pappus," the Alexandrian, of the time of Theodosius the Great, who "demonstrates it in the preface to his third book of *Mathematical Collections*." "First of all (saith he, speaking of the cells), it is convenient that they be of such figures as may cohere one to another, and have common sides, else there would be empty spaces left between them to no use but to the weakening and spoiling of the work, if anything should get in there, and therefore though a round figure be most capacious for the honey, and most convenient for the bee to creep into, yet did she not make choice of that, because then there must have been triangular spaces left void. Now, there are only three rectilineous and ordinate figures, which can serve to this purpose, and inordinate, or unlike ones, must have been, not only less elegant and beautiful, but unequal. [Ordinate figures are such as have all their ides and all their angles equal.] The three ordinate figures are triangles, squares, and hexagons; for the space about any point may be filled up either by six equilateral triangles, or four squares, or three hexagons; whereas three pentagons are too little, and three heptagons too much. Of these three, the bee makes use of the hexagon, both because it is more capacious than either of the others provided they be of equal compass, and so equal matter spent in the construction of each. And, secondly, because it is most commodious for the bee to creep into. And, lastly, because in the other figures more angles and sides must have met together at the same point, and so the work could not have been so firm and strong. Moreover, the combs being double, the cells on each side the partition are so ordered that the angles on one side insist upon the centres of the bottoms of the cells on the other side, and not angle upon or against angle; which also must needs contribute to the strength and firmness of the work."

Each cell therefore is in shape a hexagon, that is to say, a figure with six equal sides, to each of which six other hexagons attach, for each wall forms also one wall of another hexagon. The basis of each hexagonal cavity is of an obtuse three-sided pyramidal shape inverted, and consisting of three rhomboidal plates, each forming onethird of the basis of the three opposite cells; thus the edges of these three basal plates of one side support three lateral walls of three hexagons on the other side. The inverted triangular pyramid thus made by these three equal rhomboidal plates, form, at one extremity and at each pair of their posterior edges a re-entering angle, and at the other extremity a salient angle. From these edges spring the lateral walls of the hexagonal cell, this shape being superinduced by the form of the edges of the basal cavity. That the bees should have been thus guided to elect a form which combines conjunctively the advantages of strength and capacity evidently proves that it is their instinct which guides them, which, being an afflation from the highest source, ensures the most complete perfection in its result. That it cannot be the effect of simultaneous lateral pressure is proved incontestably by the whole superstructure resulting from the design of the base; and this is further corroborated by the base of one cell on one side forming invariably equal portions of the base of three cells on the opposite side, —all clearly the result of preconceived design impressed upon their sensorium. From this combination of forms results the security procured to the fragile tenement, which consists of the very smallest quantity of material that will cohere substantially, for the bees are exceedingly parsimonious of their wax, as if the production of it were attended with pain or inconvenience, and it is only upon the construction of the royal cells that a profusion of this choice material is squandered. As soon as these cohorts of bees are in active operation, it is astonishing with what pertinacity and rapidity they labour, for within the space of four-and-twenty hours they will construct a cake a foot deep and six inches wide, containing within its double area some four thousand cells. Other cakes parallel to each side of the original are being at the same time carried forward with an interval between each sufficient for two bees to pass each other *dos à dos*, and further to promote the convenience of traffic within the hive, and ready communication to its several parts, passages are left through these cakes from one to the other, so that the means of transit are opened, which of course saves much time. The queen is already making her progresses from one side of each comb to the other, and depositing her eggs as rapidly as she can, and is constantly attended by her *aides-de-camp*, as I have suggested, which act, as they evidently sometimes are, as the emissaries of her commands. They consist of ten or twelve or sometimes more, and have been previously described. They are replaced by others as they quit to obey orders, or as they retire fatigued, so that she is always surrounded. The number of eggs she will lay in a day is about two hundred. In doing this she first thrusts her head into a cell to ascertain its fitness, which having done, she withdraws it, and then curving her body she thrusts the apex of her abdomen, which tapers to the extremity for the purpose, into the cell, wherein by means of the sheaths of her curved sting, which act as an ovipositor, she places the egg at the bottom of the cell. It is possibly from some taction of this instrument that she discerns the sizes of the eggs, and thence their respective sex. This process she continues repeating, passing from one side of the comb to the other by means of the passages perforated through it, making the numbers as nearly as possible tally on each side and as opposite to each other as may be, and she will then go forward to further cakes of comb. In this way she lays about ten or twelve thousand in six weeks, depending much upon the propitiousness of the season, but the rapidity of this laying intermits according to the months; the above estimate is based upon what April and May produce, as it slackens during the summer heats and again revives in the autumn, but totally terminates with the first cold weather. She thus will lay from thirty to forty thousand or more in a year.

Apiarians do not state whether the same queen heads

another swarm on the following year, which perhaps she does in those cases of excessive fertility where her abundance is estimated at one hundred thousand, when by her sole individual capacity she populates three hives. In the more usual and ordinary case of her teeming with about seventy thousand, or fewer, she evidently heads but one swarm. With the described rapidity of the production of the cells, although the majority are store cells and not brood cells, conjunctively with her prolific laying, the population of the hive rapidly increases, which, added to the large original colony, will enable it in a propitious year to throw off a swarm of its own; but ordinarily she does not again lay drone eggs and royal eggs until the following season. The period at which to do this is taught her by the condition of the hive, as urgent for relief to its oppressive population by an exodus. The drone eggs are then laid, and are speedily succeeded by the laying of the royal eggs, so that the males of the season and the new queens may be hatched almost simultaneously, the drones slightly preceding the development of the queens. As soon as the egg of a worker is hatched, which, by means of the high temperature, is effected in four days after the laying, it, from its birth, is sedulously attended by the bees called nurse-bees. The little vermicle is very voracious and is heedfully supplied by these careful attendants, when it has consumed the quantity of bee bread already deposited in the cell by some of these nurses as soon as the egg was laid. This bee bread consists of pollen, taken from the cells by the nurses, where it is garnered for the purpose, being therein mixed with a slight quantity of honey. This, in masticating, the nurses intermingle with some secretion of their own, which gives it a sort of gelatinous frothy appearance, and upon this the young thrives so rapidly, greedily opening its jaws to receive it, that in four more days it is full grown, and fills the whole cell. The nursing-bees then cover this in with a light brown top, convex externally, and within it the larva spins for itself a cocoon to undergo its subsequent transformations. This cocoon is spun of a fine silk, which issues from the organ of the larva called the spinner, in two delicate threads, which, as they pass out, cohere together. It works at this labour for thirty-six hours, and then changes into the pupa or grub; thus it lies quiescent for three days, when it gradually undergoes its transformation into the imago, and it issues as a perfect insect about the twenty-first day after being deposited as an egg. The cocoon it has formed exactly fills the cell it has left, which still continues to serve as a brood cell until the succession of cocoons with which it is thus lined renders it too small for the purpose, it is then cleaned out by the scavengers of the hive and changed into a honey depository, but the honey stored in such a cell is never so pure as that which comes from the exclusively waxen cell. Thus is effected the transformation of the working bee, which, upon the very day of its emancipation from its nursery, commences its duties as an active member of the community, in the successive and several labours undertaken for the benefit of the commonwealth, and these it assiduously follows for the period of its natural life, which extends to about six or eight months.

The hive is now in the liveliest activity. The swarm which entered with the queen, and the large addition to the population which has already been produced from her incessant laying, are all at their several avocations. The whole hive, its entrance and the immediate vicinity, and far around is jocund with the bustle and the buzz of the busy little creatures going and coming; those returning are all laden, although some do not appear so, but these are conveying riches home within them, as they are returning from their excursions with their honeybag well filled. There is welcoming recognition at the entrance to the hive, where, on its broad platform, they all alight, and there many are to be seen touching each other with their antennæ, or refreshing themselves by the vibrations of their wings, and in doing this they often raise themselves on the hind legs, or they are resting for a few seconds before they enter. Others are to be seen arriving unrecognizable from a coloured envelope of pollen which mantles them. The incessant hum that accompanies these proceedings is like the mildest tones of the surge of the distant sea, or the inarticulate buzz of the voice of large crowds. In this seeming confusion all obey the strictest order, for each attends to his own business only; there is no collision or loss of time or labour, each one fulfilling precisely its own mission. At this period the hive is a perfect model of order, neatness, and beauty. The combs we have seen so rapidly growing are to be filled, and fresh cells are being constantly constructed. The honey there stored from the gradual gatherings of these active harvesters is partly to be reserved for the winter's needs, and is carefully husbanded, for each of these cells is, when filled, closed by a covercle of wax moulded as it is supplied to the operator in concentric circles, commencing at the edge, and each circle being completed before another is begun, and not in a spiral twist towards the

centre. To prevent the trampling of the discharging
bees from injuring the delicate structure of the walls of
the cell, each edge is furnished with a strengthening
rim of wax. The bulk of these stores is never broken,
except in bad wet seasons, in times of great dearth, or
upon any suspension of torpidity during their hiberna-
tion. For the ordinary and daily consumption of those
of the community whose labours confine them to the
hive, open stores are left. As of course it occupies
the excursions of several bees for some time to fill one
of these vases, and to prevent the liquid flowing out,
as it might do from its exceeding tenuity through the
influence of the summer heat, and the then increased
temperature of the hive, as well as from its inclined
horizontal position,—this is guarded against by the
precautional sagacity of the little creatures placing
upon it from the deposit of the very first supply a sort
of operculum, as before described, of a thicker consistency, which lies upon the top of its progressive increase,
and thus prevents its oozing. It lies upon the honey
across the transverse diameter of the cell, and consequently in a vertical position. Its purpose, like that of
the flat pieces of wood which are placed upon the water
of full pails when carried by the yoke, is to prevent its
spilling or overflowing. This small cover has to be partially removed upon the arrival of a bee with fresh store,
which she herself does by tearing aside a portion of it
to enable her to regurgitate into the cavity the portion
she has brought home; upon freeing herself from this
she does not wait to restore the dilapidation she has
caused, but proceeds on a fresh harvesting. Another
bee, whose duty it is, then readapts this cover to its
purpose, and repairs it. Their excursions to collect are

variously estimated at from one to three miles, and they make about ten a day. The bees, in their temporary distribution of labour, are something like the Indians which have caste, among whom each service has its special servitor, who never undertakes or interferes with the duties of another. The collection of pollen is almost as needful to the well-being of a hive as honey, this being used exclusively as the basis of the sustenance of the new brood in their larva state, in all their conditions of worker, drone, and queen, the perfect bee itself never partaking of it. It is variously commingled upon its application to use with secretions of their own, which convert it into bee bread or royal jelly, as the case may be, to fit it for its special employment, which is done by the nurse-bees, who diligently attend to the nurture of all the young. The cells for storing this material are not so numerous as the honey cells, and they are jotted about without any distinct order, amongst them. When a bee arrives with her store of pollen on the edge of one of these cells, she turns round with her back to it and thrusts it in as fast as she can free it from her legs, both by their aid and the twisting about of her abdomen, and then, like the honey-gatherer, commences another journey. As soon as she is gone, another bee manipulates it with a small stock of honey, and packs it closely in. Whilst all this is doing, the set which watch the condition of the hive, like surveyors, to apply repairs where necessary, or to add strength and further support to the suspended cakes of comb, impatiently await the return of the collectors of propolis; this they tear from their shanks as fast as they arrive and as quickly as they can, for it rapidly hardens, especially in fine hot weather, and they convey it away for their requirements, whilst those which collected it fly off for fresh supplies, should more be needed. Concurrently with the execution of all these things, wax is still being secreted by festoons of bees suspended wherever there is space, the sculpturer bees are still moulding cells, the queen is still laying eggs, deferentially attended, as usual, by her maids of honour; the young brood is still being fed; other bees are ventilating the hive at its entrance and within its streets and lanes by the rapid vibration of their wings; the sentinels are diligently keeping guard to repel the inimical intrusion of wasps or snails or woodlice, or the moth which is so destructive to the interior in her larva state, from the covered moveable silken retreat which she constructs impervious to the sting, and thence with impunity gets at the silk of the cocoons and consumes the wax, making, when once fairly domiciled, such fearful havoc in the hive that the bees are fain to desert it,—and the many other numerous enemies which lust for the luscious honey, or whose voracity is attracted by the poor little diligent bees themselves, but who in such contingencies exhibit invincible courage, which, if not always successful in its efforts, is always meritorious. Where self-preservation is not the prompter, or the rivalry of love the instigator, but the duration of which is limited to a season, the feuds of the animal world all seem to proceed from the urgency of their gastronomic suggestions, the acrimony of which urges craft and strength to their most powerful exhibition. To allay hunger, destruction is perpetrated and order despoiled, and thus our bees become the victims of the imperativeness of this universal law. But sometimes they are triumphant over a very large enemy; for instance, an intrusive mouse, or a slug that has slimed its way through the arched portal. They have been known to kill these enemies within the hive as they could not make them withdraw, but perplexity results from their success; they are, however, gifted with the sagacity to know that the putridity of these masses will poison with its effluvia the atmosphere of their city which no ventilation can purify, and they convert that part of their metropolis into a mausoleum, covering the carcases with a coating of propolis, alone or mixed with wax, as before noticed. Those which execute this summary martial law are the sentinels—the armed police of the hive—which guard its entrance and avenues, and patrol its streets and lanes and passages. Concurrently with all these doings, scavengers are heedfully conveying away any particles of dirt or other undesirable superfluity which may have accidentally found its way in. That all these labours produce fatigue and exact rest is proved by the circumstance that many bees are always observed in a state of repose,—perhaps only forty winks during the day just to restore exhausted energy,—for they are soon seen again to resume their toil, this inactivity never being idleness. Whether they proceed with the same kind of employment upon the renewal of their work is not known, nor how long lasts a particular kind of labour, but the change of occupation may be one of frequent occurrence, and it may be presumed that each bee severally and successively undertakes each task, that the faculty for exercising it may not be extinguished. It is very possibly a daily change, which circulates through the entire civic population of workers.

Although the labours of the bees are divided, we do not find that even the most successful observers, who have had every opportunity, by the nature of the hives they possessed, and the sagacity they applied to the detection of the most minute particulars, have been enabled to discover that these workers were permanently separated into distinct classes,—indeed, although surmising from this distribution of labour that such might be the case, and thus made alert to the discovery of its positive confirmation by direct observation, they have never been able to do so; and they strongly deny it, maintaining that these duties are individually transferable, and that they are not restricted to certain classes, already sufficiently implied by the organization of the workers. Huber, it is true, states that the wax-sculpturers—those which finish the cells to their nicety of perfection—are smaller than any of the rest of the community, to facilitate their operations within the cells, which may perhaps be a foregone conclusion.

The idea of administrative vigilance in the distribution of the labour of the community is strongly corroborated by the fact that all the labours proceed *pari passu* and in equable order, no excessive preponderance of any particular work having been observed, which would certainly sometimes be the case were there no limiting control over their individual action, and thus the harmonious concurrence of all to one effect seriously disturbed. The supposition is also strengthened by the unfailing attendance of the queen's numerous and deferential retinue, some one or other of whom, every now and then, quits that service—perhaps as an envoy on business of government—and is replaced by another. All these many circumstances lead to the presumption that the queen is the heart of the whole body, the organ which forces forward the circulation through its diverse channels, giving to all the temperate pulsation of vigorous health.

The hive is, of course, quite dark within, and to carry on the numerous operations which we have noticed are done there, either sight of a peculiar nature must lend its aid, or some faculty residing in a sensation analogous to touch, but which it may be cannot be known, nor where it may lie, but if it exist its organ is most probably the antennæ. We can, it is true, compute their eyes, which comprise more than sixteen thousand, namely, about eight thousand in each of the compound organs placed laterally upon the head, each separate eye being an hexagonal facet furnished with its separate lens and capillary branch of the optic nerve, and also edged with short hair; in this hair, therefore, may lie the particular sensation which guides them, for we cannot be sure that this large congeries of hexagonal facets facilitate sight in the dark, as in number and position they do not exceed or differ from the analogous structure and number of the same organs in many other insects which we know to be only seers by day, and which repose at night; but the hairy addition to the eyes of these bees is a structure not observed in them.

This constitution of the hive and its various operations continues during the remainder of the season until the approach of winter cautions them from venturing abroad, when, if the temperature of the hive is much lowered, they hibernate and remain in a torpid condition until the sunshine of the following spring, and with it the flowering of plants, rouses them again to resume their suspended labours. The population of the hive having continued to increase, although not so vigorously as at first, up to the very intrusion of winter, and the renewed year giving renewed energy to the queen, the population thence rapidly further increasing, it becomes inconveniently thronged, especially as spring advances and hot weather sets in. These promptings then urge her to lay drone eggs, for which preparations have already been made by the workers, who have already framed for their reception—they being much larger insects—larger cells moulded precisely in the same manner, and which are also used occasionally as receptacles for honey, and always skirt the bottom of the several combs. This task she has completed in about five days, and it is carried on precisely in the same way as is practised in the case of the neuters; and they are nurtured by nursing-workers just like them. Of these eggs she lays, as before said, about a thousand, and the workers by some instinctive faculty have framed about such a number of the needful cells. The transformations of the drone occupy about twenty-four or twenty-five days, of which three are passed in the maturing of the egg which then hatches into the larva. This occupies nearly seven days in attaining its full growth, and the remaining portion of the time is spent in its spinning its cocoon, in the same way as the larva of the worker does, and it changes into the imago. To effect all these changes in the transformations of all the sexes, a heat of about seventy degrees is indispensable, but that of the hive in summer is considerably higher. They as well as the workers are assisted to emerge from the cocoon by some of the older workers, who use their mandibles to bite through the enclosure, and who also help to cleanse them from their exuviæ.

Concurrently with the formation of the brood cells of the drones, some of the workers are constructing cells to receive the royal eggs. These cells are totally unlike the other cells of the hive, and are of a sort of pear-shape five times as large as the drone cells, and are attached laterally to the edges of the comb in a vertical position, with the narrowest part, which is the orifice, hanging downwards. In the forming of these cells the workers are very lavish of their wax, making the coats of them thick and opaque, and they are irregularly rough outside, but within very smoothly polished. Just as the construction of these cells intervenes irregularly with the formation of the cells of the drones, so does the queen intermit at intervals the laying of the drone eggs to deposit occasionally an egg in one of the royal cells, which are not usually completed at the time she commences laying them, but are finished afterwards, even during the time the larva is growing. This provision seems to be made for the earliest development of the young queens after the drones come forth, with the possible prevision that the sooner all of these young queens are fertilized that are needful for the requirements of the swarms that the hive may throw off, the sooner will the hive be rid of the incumbrance and the consumption of stores caused by the drones. The transformations of the queens take place more rapidly than the others, for in sixteen days they are completed, of which three are occupied in hatching the egg, and for five they are feeding as larvæ, and in that time attain their full growth; the cell is then closed in with a waxen cover by the workers, and the full-fed larva within is occupied in spinning its cocoon, which it takes twenty-four hours to accomplish. This cocoon is unlike that of the drones and workers, both of which completely enclose the pupa, but the royal larva only forms so much of a cocoon as will cover the head and thorax, and by which imperfection she unconsciously facilitates her destruction by her rivals in case they are permitted to attempt it before she emerges, —this being supposed to be the object of it, as the close texture of the silk of the cocoon would intercept the action of the rival queen's sting. In this state she remains in complete repose up to a part of the twelfth day, and it takes about four days more to change into the imago, which is ready to emerge on the sixteenth. In her larva state she has been very carefully and profusely supplied by her nurses with the royal jelly, made in the manner before described. This royal jelly is very stimulating, it is pungent, rather acescent, and is very different from the food supplied to the drone- and worker-larvæ. A great many of the drones being now perfect insects, some young queen, that is ready to go forth, is at length permitted to do so by her guardian protectors, for the old queen is already aware of her existence, and has more than once attempted her destruction, but from which she has been prevented. At a suitable opportunity this young queen issues, attended by a bevy of drones; she immediately ascends in a spiral direction high into the air, far out of sight, and is followed by her suitors. Their larger capacity of flight speedily permits them to overtake her, and they ascend above her; one being favoured, the rest descend again, and either at once return to the hive or frolic about in its vicinity. It is not long before this young queen returns, matured into an incipient mother. Now comes renewed hostility from her own parent, who is still prevented from the murderous assault, but who succeeds in ejecting her young rival. During this contest the hive has become a scene of confusion, and the preliminaries and accompaniments of fresh swarming take place, and in going forth she is accompanied by a large body of the present population, and thus the first swarm of the fresh season is thrown off. Other queens become gradually developed, and other swarms similarly accompany them, but each swarm successively diminishes in the number of its participating emigrants, the last consisting perhaps of not more than two thousand. The order of the hive is speedily restored after each swarming convulsion has subsided, until the population being sufficiently reduced, the motive to leave is destroyed, and the queen is then permitted to execute her murderous onslaught on the hapless young queens, which are either still embryonic, or, if developed, have not been allowed to leave their cells; but, where they have done so, and are still within the hive, her attendants and the old queen's attendants open their ranks, and the furious rivals attack each other. The contest is sharp but short, the young queen is stung to death, the body is conveyed away, and the old queen reigns paramount. Her next effort is to destroy the royal brood in their cells; the cells she tears to pieces, the young ones within, where developed, may be heard uttering a plaintive cry, whilst she sounds a triumphant note as loud as the highest note of a flute. Her throne is now free from pretenders, and after the expulsion of the drones, which then takes place, the entire harmony of the hive is restored for another season. The queen meanwhile is growing old, a new spring has set in, her stock of eggs is being exhausted, and mortality, which afflicts even royalty itself, lays her low. Now comes into operation that extraordinary faculty possessed by these insects. Her death has taken place after she had laid new spring eggs, which are to produce a further addition of neuters and a supply of drones. The loss of their queen is soon communicated to the inhabitants of the hive, confusion ensues, and labour is suspended. They group about in clusters of a dozen or more, and after about a day's intermission of the ordinary routine of labour they appear to have come to a resolution. Bustle is again renewed, and several, as the delegates of the general body, pass into the midst of the neuter brood cells, tear down the separating walls of three, kill two of the very young larvæ, convert these three cells into one by fitting alterations, and transfer the care of this vermicle to the nursing bees. Under their care, they heedfully feeding her with the royal jelly, her transformations speedily are completed, and whilst this is being done, drones are coming forth. As soon as she is ready she is aided to quit her cell. She now leaves the hive, and the drones which are already perfected accompany her; she makes her wedding tour in the air, and quickly returns as the queen-regnant of the rejoicing monarchy, whose vacant throne is again royally occupied, and the entire harmony of the hive renewed.

The quantity of pollen that is collected in the course of a season, by the diligence of the bees, has been estimated at from sixty to seventy pounds; and the weight of the honey, so affluent a hive will produce by abstraction from the bees, is calculated at as much as sometimes fifty pounds. This, however, must be vastly exceeded by the quantity collected, as it is being constantly consumed for sustenance, and for the secretion of the raw material of wax, as well as for the production of the liquid which converts this into its mouldable consistency. It is possible to estimate pretty nearly the quantity of honey required for each secretion of the raw material, by finding what the honey-bag will contain when gorged, as it is this quantity which seems to make the eight scales of it upon the ventral plates, for they cannot convey more up when they hang themselves in the festoons to secern it. But it is impossible to know what addition this liquid from their mouths makes to it when they manipulate it into its plastic state, other bees often undertaking this task, which may apply themselves to it with a larger stock than the wax-secreters possess, they being perhaps already exhausted by their labours. It is a singular fact that wax is more rapidly and largely made by feeding the bees with dissolved sugar than from the honey they collect themselves, the sugar thus evidently containing more of its productive elements.

Some of the labours within the hive are apparently continued at night, or the bees may be then revelling, after the day's toils, in social enjoyment, or otherwise more worthily employed; for, to use the words of the benevolent apiarian, the Rev. Wm. Chas. Cotton, "If you listen by a hive about nine o'clock, you will hear an oratorio sweeter than any at Exeter Hall. Treble, tenor, and bass are blended in the richest harmony. Sometimes the sound is like the distant hum of a great city, and sometimes it is like a peal of hallelujahs."

This is the history of the hive and its inhabitants. Modifications may occasionally occur, but nothing of sufficient consequence seriously to affect or neutralize this ordinary routine. It would occupy space already too largely encroached upon to go into these minute particulars, which, although parts of their general history, where treated of in special detail, are not necessarily the province of a work which speaks of them as but one member of the family of which it collectively discourses. As the space occupied by what was really essential to be known about them, has exceeded the due dimensions of their share to it, although of paramount interest, infinitely greater than that which attaches to the economy of the whole of the rest of the group combined, it will not, I trust, be considered that I terminate abruptly, in drawing here to a close.

The close of the work concurs with the termination of the history of its crowning marvel; and I take leave of my readers, with a reiteration of the hope that it may stimulate them to undertake a study, wherein, each step of their progress, expands the delightful contemplation of the manifestations of the predominance of a vast design, emanating from the paternal benevolence of an august, supreme, and wisely superintending Providence.

"To-morrow to fresh woods and pastures new."—*Milton*.

GENERAL AND GLOSSARIAL INDEX.

Abdomen, 25.
—— and its differences of form, 47.
—— causes of differences of clothing and form lie in its use, 48.
—— colour and marking and clothing of, characteristic, 47.
—— elliptical, or lanceolate and truncated, 48.
Acari infest bees, 110.
Activity of a hive at work, 348.
Acuminate, terminating gradually in a sharp point.
Affinity, doctrine of, 136.
Agassiz' 'Nomenclator Zoologicus,' 130.
Analogies between the stages of bees and flowers, 15.
Analogy, doctrine of, 138.
Andrena, general observations upon, 264.
—— geography of, 67.
—— infested by Stylops and Nomada, 208.
—— list of native species, 201.
—— natural history of, 205.
—— scientific description of, 200.
Andrenidæ, abnormal bees, 160.
—— diagram of mode of folding the tongue in repose, 39.
Animals, domestication of, 5.
Antennæ, 26, 28.
—— apparatus for cleaning, 42.
—— form and structure in Encera, 29.
—— possible complex function of, 57.

Antennæ, sexual differences in length, 233.
—— their probable use, 55, 57.
—— used as means of communication, 58.
Anthidium, general observations on, 281.
—— geography of, 75.
—— native species, 279.
—— natural history of, 282.
—— scientific description of, 279.
Anthocopa, general observations on, 292.
—— geography of, 76.
—— native species of, 292.
—— natural history of, 293.
—— scientific description of, 290.
Anthophora, general observations on, 238.
—— geography of, 70.
—— infested by Melecta, 240.
—— list of native species, 238.
—— natural history of, 238.
—— scientific description of, 236.
—— trophi of, 29.
Apathus, general observations on, 304.
—— geography of, 77.
—— list of native species, 304.
—— scientific description of, 302.
—— the Bombi they infest, 306.
Apidæ, diagram of the mode of folding the tongue in repose, 39.
—— = normal bees, 160, 227.
Apis, general observations on, 321.
—— geography of, 79.

Apis, native species, 321.
—— natural history of, 322.
—— origin of names, 321.
—— scientific description of, 318.
—— see "Bee" and "Bees."
Appearance of bees intermittent, 54.
Appendiculated, when there is a small appendage, as in the lip of Halictus, and at the end of the marginal cell of the wings, etc.
Arrangement and description of British bees, 184.
Artesian well, peculiar results from its soil, 223.
Articulate, where jointed, or the point of attachment.
Artisan bees = Dasygasters, 272.
Aryans, one of the primitive divisions of the human race, 4.
Atmosphere, its conditions affect bees, 50.
Aulacus, 287.
Auriculated, with a small ear-like appendage.

Bee, constructive habits of the, early noticed, 93.
—— general history of the, 17.
—— parasites, 115.
—— parasitism limited, 264.
—— probably earlier known to man than the silkworm, 6.
—— Queen, description of, 322.
—— see "Apis."
—— several species of, 87.
—— symbol of royalty with the Egyptians, 5.
—— The, one of the Suras of the Koran, 90.
—— why attractive, 1.
Bee-bread, 347.
Bees, amount of their susceptibility of pain, 57.
—— construction of cells, 327.
—— duties performed in the hive, 325.
—— duties transferable, 336.
—— early cultivated, 3, 90, 91.
—— economy, early known, 92.
—— emit an odour, 52.

Bees enemies, 51.
—— extent of flight, 340.
—— flight, modes of, varies, 49.
—— found in the Orkneys, 7.
—— genera of, determined by an artificial mode, 170.
—— habits of, in America, 7.
—— hairiness of, reason of, 14.
—— intimately connected with flowers, 3.
—— largely contribute to the impregnation of plants, 11.
—— make about ten journeys a day, 351.
—— many disclosed in autumn for the following year's spring flight, 53.
—— not early risers, 51.
—— number of eyes, 355.
—— other than social, also known, 8.
—— rarely walk, 50.
—— sagacity in finding the honey of flowers, 13.
—— scientific arrangement and description of the genera of, 184.
—— secretion of wax, 325.
—— stages of life of,—egg, 18.
—— —— larva, 19.
—— —— pupa, 22.
—— —— imago, 23.
—— swarming, 337.
—— their relative perfection, 56.
—— voice, a scale of music, 49.
Beehive represented on a tomb at Thebes, 6.
Beehives moved on rafts, 84.
Bifid, divided into two parts.
Binomial system invented by Linnæus, 129.
Body of the bee, its structure, 25.
Bombus, difficulty in determining the species of the males, 311.
—— general observations on, 310.
—— geography of, 78.
—— infested by *Apathus*, 311.
—— list of native species, 308.
—— natural history of, 312.
—— peculiarities in times of appearance, 312.
—— scientific description of, 307.

Boss of mesothorax, 45.
Bougie, derivation of, 84.
British bees, new arrangement of, 153, 158.

Carder bees, 316.
Carelessness of describers of new species, 125.
Carinated, having a longitudinal elevated line.
Carpenter bees, 286.
Cells of hive, geometrical form of, 343.
—————— results from instinct, 343.
—— how constructed, 342.
—— of wings characteristic, 44.
Cenobites = social bees, 167, 302.
Ceratina, disputed parasitism of, 247.
—— general observations on, 246.
—— geography of, 71.
—— list of native species, 246.
—— natural history of, 247.
—— scientific description of, 245.
Cereal plants early cultivated, 4.
Chelostoma, general observations on, 286.
—— geography of, 76.
—— infested by Fœnus, 287.
—— native species of, 285.
—— natural history of, 286.
—— scientific description of, 283.
Chrysis infests Chelostoma, 287.
—— infests Halictus, 219.
—— infests Osmia, 302.
Cibarial apparatus = trophi = collective organs of the mouth, 163.
Cilissa, general observations on, 213.
—— geography of, 67.
—— list of native species, 213.
—— scientific description of, 211.
Clavate, club-shaped.
—— antennæ, 28.
Claws, 42.
—— reflected, 285.
Climate inoperative on low forms of life, 24.
Clothing of bees, 60.
Clypeus, 26, 28.

Coadunate, closely united without perceptible articulation.
Cœlioxys, difficulty of their specific separation, 267.
—— general observations on, 267.
—— geography of, 74.
—— list of native species, 267.
—— parasitical on Megachile and Saropoda, 267.
—— scientific description of, 265.
Collar, 41.
Colletes, general observations on, 187.
—— geography of, 64.
—— list of native species, 187.
—— natural history of, 187.
—— parasites upon, 190.
—— scientific description of, 185.
Colour of bees, 60.
—— more intense in males than females, 52.
—— most conspicuous in parasites, 66, 105.
Combs, structure of, 315.
Corbiculum, 319.
Correlative relations of structure and function, 10.
Cotton, Rev. Chas. Wm., a distinguished apiarian, 361.
Coxa, or hip, 41.
—— useful as a specific character, 42.
Compound eyes, 26, 27.
Compressed, when the transverse section is shorter than the vertical.
Constricted, with tightened edges.
Conterminous, where the joints follow each other in a straight line of succession.
Crenulated, cut into segments of very small circles.
Cubital cells of wings, 45.
Cuckoo bees, = Nudipedes, 249.
'Cui bono?' answer to, 141.
Curtis, inferior merit of his system, 152.

Dasygasters, artisan bees, 167, 269.
Dasypoda, general observation on, 225.

Dasypoda, geography of, 69.
—— native species, 225.
—— natural history of, 226.
—— scientific description of, 224.
Deflected, when bent downwards.
Dentate, toothed.
Depressed, when the vertical section is shorter than the transverse.
Describers, duties of, 125.
Describing, modes of, before Linnæus, 129.
Differences of appearance between the parasite and the sites, 260.
Digiti, anterior tarsi, 42.
Dissimilarity frequent between the sexes, 52.
Domestication of animals, 5.
Dorylus, 311.
Drone = male bee, description of, 323.

Edentate, without teeth.
Egg of bees, 18.
Egyptian hieroglyphics and sculptures represent the bee, 6.
Elenchus, habits of, described by Dale, 113.
—— infests Halictus, 113, 219.
Elliptical, oval but with the longitudinal diameter more than twice the length of the transverse.
Enemies of bees, 51.
Epeolus, general observations on, 260.
—— geography of, 73.
—— native species, 260.
—— parasitical on Colletes, 190–260.
—— scientific description of, 258.
Epipharynx, 29, 30.
Eucera, general observations on, 232.
—— geography of, 70.
—— infested by Nomada sexcincta, 235.
—— native species, 232.
—— natural history of, 234.
—— scientific description of, 231.

Face of bees, 26, 27.
Families, characteristics of, differ, 136.
Family, 134.
Feeling of bees, 56.
Femur, or thigh, 41.
Fertilization of flowers produced by bees, 11, 51.
Feuds of animals, the occasion of, 352.
Filiform, thread-like, of uniform thickness.
—— antennæ, 28.
Fimbriated, = fringed.
Flagellum of antennæ, 18.
Flight of bees, variation of their modes, 49.
Floral clock of Linnæus, 50.
Flowers, the, chiefly agreeable to bees, 15.
—— earliest, sought by the bees, 14.
—— fertilized by bees, 11, 51.
Fœnus infests Chelostoma, 287.
Forcipate, when crossing each other.
Foreign bees, conspicuous genera of, 101.
Form of parasitical bees often adapted to that of their sites, 48.
—— determined by function, 48.
Fossorial Hymenoptera, 45.
Fruit preserved in honey, 83.
Fusiform, = spindle-shaped.

Genæ, 26, 28.
Genera of bees determined artificially, 176.
—— that emit scents, 296.
—— with and without parasites, 264.
Geniculated, bent like a knee or angle.
Genus, 132.
—— type of, 133.
Geography of the British genera of bees, 61.
Gibbous, = irregularly swollen.
Glabrous, without hair or pubescence.
Gregarious, its application to bees, 57.

GENERAL AND GLOSSARIAL INDEX. 367

Habit, 127.
Habitat, 127.
Habits, 127.
—— and structure correlative, 24.
Halictophagus, 115.
Halictus, general observations on, 216.
—— geography of, 68.
—— its enemies, 220.
—— list of native species, 215.
—— natural history of, 217.
—— parasites that infest it, 219.
—— peculiar autumnal appearance, 218.
—— scientific description of, 214.
—— structure of labrum, 30.
Hastate, halberd shaped.
Head of bees, 26.
Hedychrum infests Halictus, 219.
Heriades, general observations on, 288.
—— geography of, 76.
—— native species of, 288.
—— scientific description of, 288.
Hindoo Koosh, supposed cradle of the human race, 3.
Hirsute, covered with long stiffish hairs, thickly set.
Hives, darkness of, 355.
—— moved on rafts, 85.
Homer mentions bees, 6.
Honey, different kinds of, 87.
—— green, 87.
—— its use in the East, 83.
—— mode of lapping, described by Réaumur, 35.
—— mode of storing, 350.
—— prescribed by Mahomet, 91.
—— quantity in a well-filled hive, 360.
—— sometimes poisonous, 86.
—— used in medicine by the Egyptians, 90.
Honey-bee, see "Apis," "Bee," "Bees."
———————— mode of secreting wax, 330.
Hypopharynx, 29.

Imago of bees, 23.

Inosculation, point of close contact or attachment.
Insect-feeding reptiles before glacial period, 5.
Inserted, where joined.
Instinct, its applications, 56.
—— occasional divergence of, 55.
—— of bees, 55.

Job mentions bees, 6.

Kirby's merits, 144.
—— system of bees, 147.

Labial palpi, 30, 32.
—— number of joints invariable, 32.
—— structure in Andrenidæ, 32.
—— structure in Apidæ, 32.
Labium = lower lip, 30, 31.
Labrum = upper lip, 28, 30.
Lacerate, with a roughened irregular edge.
Lanceolate, oblong but gradually tapering.
Latreille's classification not adopted, 168.
Leg, diagram of, 42.
Legs, general description of, 41.
Length of an insect is taken from the front of the head to the apex of the abdomen; the breadth, or the expansion of the wings, it is not usual to give, excepting under such circumstances as would be particularly mentioned, viz. in cases of an excessive enlargement or diminishment of the typical size.
Life, duration of, of bees, 54.
Line, the twelfth part of an inch; the ordinary measure used in entomology for the fractions of an inch, unless the insect is much more than an inch long.
Linnæus, author of the binomial system, 129.
—— great merits of, 129.
Lobated, divided into equal rounded parts.

Low forms of life unaffected by climate, 24.
Lunate, semicircular.
Lunulate, crescent-shaped.

Macropis, general observations on, 222.
—— geography of, 68.
—— native species, 221.
—— scientific description of, 220.
—— strong analogy to the Scopulipedes, 222.
Maculæ indicantes, 13.
Mahomet prescribes honey, 91.
Males, how to be united to their partners, 179.
Mandibles, 30, 40.
—— used for boring, 44.
Marginal cells of wings, 45.
Marginate, edged with a ridge.
Mason bees, 296.
Maxillæ, 30, 31.
Maxillary palpi, 30, 32.
—— number of joints invariable in Andrenidæ, 32.
—— number of joints variable in the Apidæ, 32.
Megachile, general observations on, 272.
—— geography of, 74.
—— infested by Cœlioxys, 275.
—— list of native species, 271.
—— natural history of, 273.
—— scientific description of, 269.
Melecta, general observations on, 255.
—— geography of, 72.
—— list of native species, 255.
—— parasitical on Anthophora, 240.
—— scientific description of, 255.
—— very pugnacious, 258.
Melittobia, a parasite upon Anthophora, 241.
Meloë proscarabæus, parasitical on bees, 110.
—— said to infest Andrena, 209.
Mesothorax, 26, 44.
Metallic colouring of bees, 248.
Metathorax, 26.
Metropolis, 128.

Miltogramma, parasitical upon Colletes, 190.
Mode of killing coloured insects, 253.
Moniliform, bead-like.
—— antennæ, 129.
Monodontomerus, parasitical on Anthophora and Osmia, 302.
Moths help to fertilize flowers, 13.
Motives for new arrangement, 163.
Mouth, organs of = trophi = cibarial apparatus, 163.
Mucronated, having one or more short stout processes.
Mutilla, parasitical on bees, 117.

Names usually given from a sexual peculiarity, 232.
Natural history, attractions of, 141.
—— modes of treating, 140.
Natural system, 139.
Nature, its large operations, 8.
Nectaria of plants indicated to bees by a difference of colour, 12.
Nervures of wings, 44.
Nomada, general observations on, 252.
—— geography of, 72.
—— intermittent appearance of N. Fabriciana, 230.
—— list of native species, 250.
—— scientific description of, 249.
—— sexcincta infests Eucera, 235.
—— the bees infested by them, 253.
Nomenclature simplified by Linnæus, 130.
Nudipedes, = cuckoo-bees or parasites, 116, 167, 249.
Nylander's mode of determining the species of Cœlioxys, 268.

Obsolete, more or less inapparent.
Ocelli = simple eyes = stemmata, 26, 27.
Oman, no bees in the province of, 84.
Osmia, general observations on, 296.
—— geography of, 76.
—— list of native species, 295.

Osmia, natural history of, 296.
—— parasites of, 302.
—— scientific description of, 294.
Ovate, oval, but with the ends circumscribed by unequal segments of circles.
Ovipositor = egg-depositor, 17.

Pain, doubtful susceptibility of, 57.
Palmæ, 41.
Palmated, spread like a hand.
Palpi, their probable use, 55.
Panurgus, general observations on, 229.
—— geography of, 69.
—— infested by Nomada Fabriciana, 230.
—— list of native species, 228.
—— natural history of, 229.
—— scientific description of, 227.
Paraglossæ, 33.
—— obsolete in the artisan bees, 33.
—— where attached, 33.
Parasites, different kinds of, 110.
—— of bees, 109.
Parasitical bees always the most highly coloured, 66, 105.
—— unlike the sitos, 116.
—— Cenobites, 302.
Passions of bees, 56.
Pecten or comb, a fringe of very short stiff hair attached to an organ, for various purposes.
Pectinated, having an edge like a comb.
Pediculus Melittæ, 209.
Petiole, a foot-stalk.
Pharynx, 29, 30.
Pile, long loose hair.
Pilose, with long, distinct, flexible hair.
Plantæ, 42, 46.
—— structure of, in hive-bee, 46.
Plants agreeable to bees, 15.
—— impregnated by bees, 11.
Pleasures attending the pursuit of natural history, 14.
Plumose, with long hair, but not thick.
Pollen, collection of, 351.

Pollen, mode of collecting and transferring from limb to limb, 43.
—— probable reasons for the ways of carrying, 47.
—— quantity usually collected, 360.
Polliniferous, = pollen-collecting.
Posterior legs, their structure for the conveyance of pollen, 46.
—— - where attached, 46.
Post-scutellum, 26, 45.
Priority, law of, the basis of synonymy, 131.
Proboscis, 39.
Process, a protuberance.
Processes in bees, peculiarities of, 258.
Propolis, nature of, 340.
Prosopis emits an agreeable odour, 195.
—— general observations upon, 193.
—— geography of, 65.
—— list of native species, 192.
—— presumed parasitism of, 193.
—— scientific description of, 191.
—— supposed liable to Stylops, 195.
Prothorax, 26, 41.
Pubescent, covered with short fine hair.
Pubescent, hirsute, setose, pilose, plumose, various relative conditions of hairiness.
Pulvillus, 42.
Punctate, impressed with many points.
Punctulate, with fine impressed points.
Punctured, with coarsely impressed points.
Pupa of bees, 22.

Queen-bee, administrative function of, 336.
—— and worker constitute a unity, 331.
—— description of, 322.
—— etiquette of attendants, 329.

Queen-bee, great fertility of, 334.
—— loss of, disorganizes the hive, 335.
—— number of eggs laid by, 346.

Ray's merits, 142.
Réaumur's description of the mode of lapping honey, 35.
—— description of the structure of the tongue, 35.
Recurrent nervures of wings, 45.
Retuse, with an obtuse cavity.
Ridged, with a slight projecting margin.
Rugose, rough or irregularly wrinkled.

St. Fargeau's merits, 151.
Sanskrit notice of bees and honey, 92.
Saropoda, general observations, 243.
—— geography of, 71.
—— native species of, 243.
—— rapidity of flight, 245.
—— scientific description of, 242.
—— vivacity of its eyes, 244.
Scape of antennæ, 28.
Scent emitted by bees, 52.
Scientific arrangement and description of the genera, 184.
———— principles of, 118.
—— cultivation of British bees, 142.
Scopulipedes = brush-legged bees, 163, 227.
Sculpture, 60.
Scutellum, 26, 45.
Senses of bees, 56.
Sensorium of bees, 55.
Serrate, edged like a saw.
Serratulate, edged like a fine saw.
Setæ, slightish bristles.
Setiform, like bristles.
Setose, bristled.
Shakespeare on the polity of the bee, 1.
Shemitic branch of the human race, 4.
Sight of bees, 56.

Simple eyes = ocelli = stemmata, 26, 27.
Sinus, a cavity.
Sitos, the supporter of a parasitical bee.
Sizes, differences of, what caused by, 41.
Smell of bees, 56.
Social bees, = Cenobites, 302.
Species, 122.
—— name of, 128.
—— the basis of natural science, 121.
—— vary in number of individuals, 123.
Specific character, 124.
—— descriptions, 125.
—— differences, 123.
Sphecodes, difficulty of specific distinction in, 198.
—— doubts as to its parasitism, 199.
—— general observations on, 197.
—— geography of, 66.
—— list of native species, 197.
—— scientific description of, 196.
Spines at apex of abdomen of bees, 268.
Spinose, with minute spiny processes.
Spinulose, with fine spiny processes.
Spiral hair of the scopa, 226, 229.
Spurs, 42.
Squamulæ = epaulettes = wing-scales, 26, 44.
State of Great Britain before the glacial period, 5.
Stelis, general observations on, 263.
—— geography of, 73.
—— infests Osmia, 302.
—— list of native species, 263.
—— scientific description of, 262.
Stemmata = simple eyes = ocelli, 26, 27.
Stephens, inferior merit of his system, 152.
Strepsiptera parasitical on bees, 111.
Strigilis, 42.

GENERAL AND GLOSSARIAL INDEX. 371

Structure and habits correlative, 24.
—— of the body of the bee, 25.
—— similarity of, caused by direct and proximate affinities, 48.
Stylops infests Andrena, 208.
—— infests Halictus, 219.
—— Kirby's description of, 112.
—— manners of, described by Thwaites, 114.
—— some particulars of its history, 208.
Sub, a prefix indicating the diminution of a condition, as subhastate, subovate, subtruncate, etc. etc.
Submarginal cells of wings, 45.
Swarming, 358.
Synonymy, 130.
System, value of, 119.

Tarsus of fore legs in some males greatly dilated, 43.
—— or foot, 41.
Taste of bees, 56.
Thorax, 26, 41.
Tibia, or shank, 41.
Tomb at Thebes with representation of a beehive, 6.
Tongue improperly called labium, 34.
—— of Andrenidæ folded in repose, 39.
—— of Apidæ folded in repose, 39.
—— once thought tubular, 34.
—— where situated, description of it, 33.
Topical geography of British bees, 96.
Tooth, a long sharp process.
Toothed, spinose, spinulose, tuberculated, mucronated, dentate, the various conditions of extraneous prominences or processes.
Transformations of worker bee, 347.
——————— of the drone, 356.
——————— of the Queen, 357.
Transverso-cubital nervures of wings, 45.

Travellers, suggestions to, 64, 95.
Trifid, divided into three parts.
Trivial name, 128.
Trochanter, 41.
Trophi = organs of the mouth, 26, 29.
—— diagram of, 30.
Truncated, abruptly terminated.
Tuberculated, with small processes.
Turonian branch of the human race, 4.

Uses of bees in the impregnation of plants, 11.

Vedas mention bees, 6.
Velum, 42.
Ventilation of the hives, 328.
Ventral segments, peculiarities of structure of, 234.
Vernacular names of insects, 9.
Vertex, 26.
Vertigo of bees, 87.
Voice of bees, 49.

Wagtails destroy fossorial Hymenoptera, 306.
Wax, secretion of, 325.
Wax used by the Romans, 85.
Westwood's classification not adopted, 168.
Wild bees, 8.
—— come forth early in the spring, 10.
Will of bees, 56.
Willughby's merits, 143.
Wing, treatise on the, 45.
Wing-hooklets for uniting the upper and lower wings, 45.
Wing-scales = squamulæ, 26.
Wings, 44.
—— diagram of, 45.
Worker-bee, description of, 324.
———— duties performed by, 325.
———— peculiarities of structure, 330.
———— secretion of wax, 325.

Xenophon's description of poisonous honey, 86.

PLATE I.

1 ♂. Colletes Daviesiana, *male.*
1 ♀.   ,,   ,,   *female.*
2 ♂. Prosopis dilatata, *male.*
2 ♀. Prosopis signata, *female.*
3 ♂. Sphecodes gibbus, *male.*
3 ♀.   ,,   ,,   *female.*

PLATE II.

1 ♂. Andrena fulva, *male*.
1 ♀.   ,,   ,, *female*.
2 ♂. Andrena cineraria, *male*.
2 ♀.   ,,   ,, *female*.
3 ♂. Andrena nitida, *male*.
3 ♀.   ,,   ,, *female*.

PLATE III.

1 ♂. Andrena Rosæ, *male*.
1 ♀.   ,,   ,, *female*.
2 ♂. Andrena longipes, *male*.
2 ♀.   ,,   ,, *female*.
3 ♂. Andrena cingulata, *male*.
3 ♀.   ,,   ,, *female*.

PLATE IV.

1 ♂. Halictus xanthopus, *male*.
1 ♀.   „   „   *female*.
2 ♂. Halictus flavipes, *male*.
2 ♀.   „   „   *female*.
3 ♂. Halictus minutissimus, *male*.
3 ♀.   „   „   *female*.

PLATE V.

1 ♂. Cilissa tricincta, *male*.
1 ♀. ,, ,, *female*.
2 ♂. Macropis labiata, *male*.
2 ♀. ,, ,, *female*.
3 ♂. Dasypoda hirtipes, *male*.
3 ♀. ,, ,, *female*.

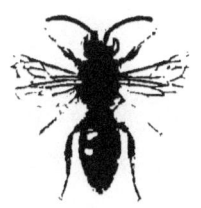

PLATE VI.

1 ♂. Panurgus Banksianus, *male.*
1 ♀.   ,,   ,, *female.*
2 ♂. Eucera longicornis, *male.*
2 ♀.   ,,   ,, *female.*
3 ♂. Anthophora retusa, *male.*
3 ♀.   ,,   ,, *female.*

PLATE VII.

1 ♂. Anthophora furcata, *male*.
1 ♀.   ,,   ,, *female*.
2 ♂. Saropoda bimaculata, *male*.
2 ♀.   ,,   ,, *female*.
3 ♂. Ceratina cærulea, *male*.
3 ♀.   ,,   ,, *female*.

PLATE VIII.

1 ♂. Nomada Goodeniana, *male*.
1 ♀.   ,,   ,,   *female*.
2 ♂. Nomada Lathburiana, *male*.
2 ♀.   ,,   ,,   *female*.
3 ♂. Nomada sexfasciata, *male*.
3 ♀.   ,,   ,,   *female*.

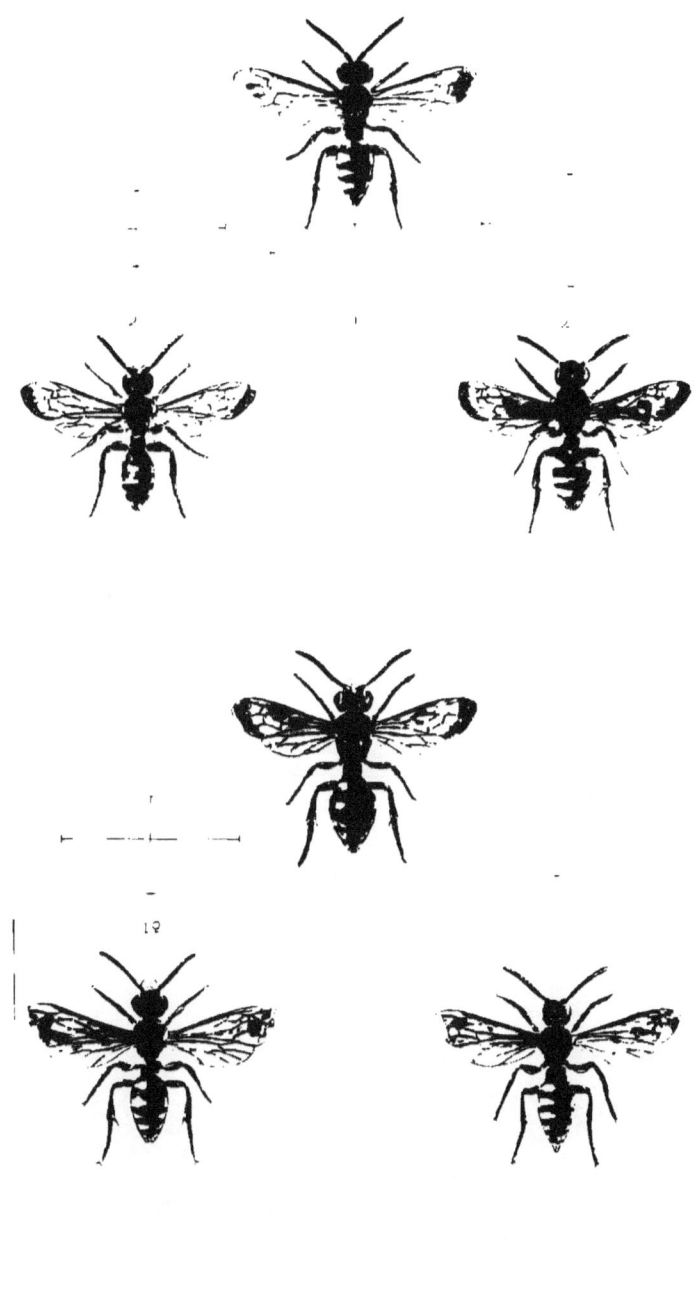

PLATE IX.

1 ♂. Nomada signata, *male*.
1 ♀.   „   „   *female*.
2 ♂. Nomada Fabriciana, *male*.
2 ♀.   „   „   *female*.
3 ♂. Nomada flavoguttata, *male*.
3 ♀.   „   „   *female*.

PLATE X.

1 ♂. Nomada Jacobææ, *male*.
1 ♀.   ,,   ,,   *female*.
2 ♂. Nomada Solidaginis, *male*.
2 ♂* (should be ♀). ,, *female*.
3 ♂. Nomada lateralis, *male*.
3 ♀.   ,,   ,,   *female*.

PLATE XI.

1 ♂. Melecta punctata, *male*.
1 ♀. ,, ,, *female*.
2 ♂. Epeolus variegatus, *male*.
2 ♀. ,, ,, *female*.
3 ♂. Stelis phæoptera, *male*.
3 ♀. ,, ,, *female*.

PLATE XII.

1 ♂. Cœlioxys Vectis, *male*.
1 ♀.   ,,   ,, *female*.
2 ♂. Megachile maritima, *male*.
2 ♀.   ,,   ,, *female*.
3 ♂. Megachile argentata, *male*.
3 ♀.   ,,   ,, *female*.

PLATE XIII.

1 ♂. Anthidium manicatum, *male.*
1 ♀.   ,,   ,,   *female.*
2 ♂. Chelostoma florisomne, *male.*
2 ♀.   ,,   ,,   *female.*
3 ♂. Heriades truncorum, *male.*
3 ♀.   ,,   ,,   *female.*

PLATE XIV.

1 ♂. Osmia bicolor, *male*.
1 ♀. ,,   ,,   *female*.
2 ♂. Anthocopa Papaveris, *male*.
2 ♀. ,,   ,,   *female*.
3 ♂. Osmia leucomelana, *male*.
3 ♀. ,,   ,,   *female*.

PLATE XV.

1 ♂. Apathus rupestris, *male*.
1 ♀. ,, ,, *female*.
2 ♂ (should be ♀). Apathus campestris, *female*.
2 ♀. Apathus vestalis, *female*.
3 ♀. Bombus fragrans, *female*.
4 ♂. ,, Soroensis (var. Burrellanus), *male*.

PLATE XVI.

1 ♀. Bombus Harrisellus, *female*.
2 ♀.   ,,   Lapponicus, *female*.
3 ♀.   ,,   sylvarum, *female*.
4 ♂. Apis mellifica, *male*.
4 ♀.   ,,   ,,   *female*.
4 ☿.   ,,   ,,   *neuter*.

LOVELL REEVE & CO.'S

PUBLICATIONS IN

Botany, Conchology, Entomology,

CHEMISTRY, TRAVELS, ANTIQUITIES,

ETC.

"None can express Thy works but he that knows them;
And none can know Thy works, which are so many
And so complete, but only he that owes them."
George Herbert.

LONDON:
LOVELL REEVE AND CO., 5, HENRIETTA STREET, COVENT GARDEN.
1866.

CONTENTS.

	PAGE
BOTANY	3
FERNS AND MOSSES	9
SEAWEEDS	10
FUNGI	11
SHELLS AND MOLLUSKS	12
INSECTS	14
TRAVELS	15
ANTIQUARIAN	16
MISCELLANEOUS	17
WORKS IN PREPARATION	19

All Books sent post-free to any part of the United Kingdom on receipt of a remittance for the published price.
Post-Office Orders to be made payable at KING STREET, COVENT GARDEN.

LIST OF WORKS

PUBLISHED BY

LOVELL REEVE & CO.

BOTANY.

HANDBOOK OF THE BRITISH FLORA; a Description of the Flowering Plants and Ferns indigenous to, or naturalized in, the British Isles. For the Use of Beginners and Amateurs. By GEORGE BENTHAM, F.R.S., President of the Linnean Society. Crown 8vo, 680 pp., 12s.

Distinguished for its terse and clear style of description; for the introduction of a system of Analytical Keys, which enable the student to determine the family and genus of a plant at once by the observation of its more striking characters; and for the valuable information here given for the first time of the geographical range of each species in foreign countries.

HANDBOOK OF THE BRITISH FLORA, ILLUSTRATED EDITION; a Description (with a Wood-Engraving, including dissections, of each species) of the Flowering Plants and Ferns indigenous to, or naturalized in, the British Isles. By GEORGE BENTHAM, F.R.S., President of the Linnean Society. Demy 8vo, 2 vols., 1154, pp. 1295 Wood-Engravings, from Original Drawings by W. Fitch. £3. 10s.

An illustrated edition of the foregoing Work, in which every species is accompanied by an elaborate Wood-Engraving of the Plant, with dissections of its leading structural peculiarities.

THE FIELD BOTANIST'S COMPANION; a Familiar Account, in the Four Seasons, of the most common of the Wild Flowering Plants of the British Isles. By THOMAS MOORE, F.L.S. One volume, Demy 8vo, 424 pp.. With 24 Coloured Plates, by W. FITCH, 21s.

An elegantly-illustrated volume, intended for Beginners, describing the plants most readily gathered in our fields and hedge-rows, with the progress of the seasons. Dissections of the parts of the flowers are introduced among the Figures, so that an insight may be readily obtained not only of the Species and name of each plant, but of its structure and characters of classification.

THE FLORAL MAGAZINE, containing Figures and Descriptions of New Popular Garden Flowers. By the Rev. H. Honywood Dombrain, A.B. Imperial 8vo. Published Monthly, with 4 Plates, 2s. 6d. coloured. Vol. I. to IV., each with 64 plates, £2. 2s.

Descriptions and Drawings, beautifully coloured by hand, of new varieties of Flowers raised by the nurserymen for cultivation in the Garden, Hothouse, or Conservatory.

THE TOURIST'S FLORA; a Descriptive Catalogue of the Flowering Plants and Ferns of the British Islands, France, Germany, Switzerland, Italy, and the Italian Islands. By JOSEPH WOODS, F.L.S. Demy 8vo, 504 pp., 18s.

Designed to enable the lover of botany to determine the names of any wild plants he may meet with while journeying in our own country and the countries of the Continent most frequented by tourists. The author's aim has been to make the descriptions clear and distinct, and to comprise them within a volume of not inconvenient bulk.

CONTRIBUTIONS TO THE FLORA OF MENTONE. By J. TRAHERNE MOGGRIDGE. Royal 8vo. Parts I. and II., each, 25 Coloured Plates, 15s.

In this work a full page is devoted to the illustration of each Species, the drawings being made by the author from specimens collected by him on the spot, and they exhibit in vivid colours the beautiful aspect which many of our wild flowers assume south of the Alps.

A FLORA OF ULSTER, AND BOTANIST'S GUIDE TO THE NORTH OF IRELAND. By G. DICKIE. M.D., F.L.S., Professor of Botany in the University of Aberdeen. A pocket volume, pp. 176, 3s.

A small volume, not exclusively of local interest, containing, as it does, much valuable information relative to the geographical and altitudinal range of the Species.

VICTORIA REGIA; or, Illustrations of the Royal Water Lily, in a series of Figures chiefly made from Specimens flowering at Syon and at Kew, by W. FITCH, with Descriptions by Sir W. J. HOOKER, F.R.S. Elephant folio, 21s.

A superb series of illustrations of this wonderful plant, with an elaborate series of dissections.

CURTIS'S BOTANICAL MAGAZINE, comprising the
Plants of the Royal Gardens of Kew, and of other Botanical Establishments. By Dr. J. D. HOOKER, F.R.S., Director of the Royal Gardens. Royal 8vo. Published Monthly, with 6 Plates, 3s. 6d. coloured. Vol. XXII. of the Third Series (being Vol. XCII. of the entire work) in course of publication. A Complete Set from the commencement may be had.

Descriptions and Drawings, beautifully coloured by hand, of newly-discovered plants suitable for cultivation in the Garden, Hothouse, or Conservatory.

THE RHODODENDRONS OF SIKKIM-HIMALAYA;
being an Account, Botanical and Geographical, of the Rhododendrons recently discovered in the Mountains of Eastern Himalaya, from Drawings and Descriptions made on the spot, by Dr. J. D. Hooker, F.R.S. By Sir W. J. HOOKER, F.R.S. Folio, 30 Coloured Plates, £3. 16s.

Illustrations on a superb scale of the new Sikkim Rhododendrons, now being cultivated in England, accompanied by copious observations on their distribution and habits.

MONOGRAPH OF ODONTOGLOSSUM, a Genus of the
Vandeous Section of Orchidaceous Plants. By JAMES BATEMAN, Esq., F.R.S. Imperial folio. Parts I. to III., each with 5 Coloured Plates, and occasional Wood Engravings, 21s.

Designed for the illustration, on an unusually magnificent scale, of the new and beautiful plants of this favoured genus of *Orchidacea*, which are being now imported from the mountain-chains of Mexico, Central America, New Granada, and Peru.

SELECT ORCHIDACEOUS PLANTS. By ROBERT
WARNER, F.R.H.S. With Notes on Culture by B. S. WILLIAMS. In Ten Parts, folio, each, with 4 Coloured Plates, 12s. 6d.; or, complete in one vol., cloth gilt, £6. 6s.

PESCATOREA. Figures of Orchidaceous Plants, chiefly
from the Collection of M. PESCATORE. Edited by M. LINDEN, with the assistance of MM. G. LUDDEMAN, J. E. PLANCHON, and M. G. REICHENBACH. Folio, 48 Coloured Plates, cloth, with morocco back, £5. 5s., or whole morocco, elegant, £6. 6s.

GENERA PLANTARUM, ad Exemplaria imprimis in Herbariis Kewensibus servata definita. By GEORGE BENTHAM, F.R.S., President of the Linnean Society, and Dr. J. D. HOOKER, F.R.S., Assistant-Director of the Royal Gardens, Kew. Vol. I. Part I. pp. 454. Royal 8vo, 21s. Part II., 14s.

This important work comprehends an entire revision and reconstruction of the Genera of Plants. Unlike the famous Genera Plantarum of Endlicher, which is now out of print, it is founded on a personal study of every genus by one or both authors. The First Part contains 56 Natural Orders and 1287 Genera. The Second, now printing, will contain as many more. The whole will be completed in Four or Five Parts.

FLORA OF THE ANTARCTIC ISLANDS; being Part I. of the Botany of the Antarctic Voyage of H.M. Discovery Ships 'Erebus' and 'Terror,' in the years 1839–1843. By Dr. J. D. HOOKER, F.R.S. Royal 4to. 2 vols., 574 pp., 200 Plates, £10. 15s. coloured. Published under the authority of the Lords Commissioners of the Admiralty.

The 'Flora Antarctica' illustrates the Botany of the southern districts of South America and the various Antarctic Islands, as the Falklands, Kerguelen's Land, Lord Auckland and Campbell's Island, and 1370 species are enumerated and described. The plates, which are executed by Mr. Fitch, and beautifully coloured, illustrate 370 species, including a vast number of exquisite forms of Mosses and Seaweeds.

FLORA OF NEW ZEALAND; being Part II. of the Botany of the Antarctic Voyage of H.M. Discovery Ships 'Erebus' and 'Terror,' in the years 1839–1843. By Dr. J. D. HOOKER, F.R.S. Royal 4to, 2 vols., 733 pp., 130 Plates. £16. 16s. coloured. Published under the authority of the Lords Commissioners of the Admiralty.

The 'Flora of New Zealand' contains detailed descriptions of all the plants, flowering and flowerless, of that group of Islands, collected by the Author during Sir James Ross' Antarctic Expedition; including also the collections of Cook's three voyages, Vancouver's voyages, etc., and most of them previously unpublished. The species described amount to 1767; and of the Plates, which illustrate 313 Species, many are devoted to the Mosses, Ferns, and Algæ, in which these Islands abound.

FLORA OF TASMANIA; being Part III. of the Botany of the Antarctic Voyage of H.M. Discovery Ships 'Erebus' and 'Terror,' in the years 1839–1843. By Dr. J. D. HOOKER, F.R.S. Royal 4to, 2 vols., 972 pp., 200 Plates, £17. 10s., coloured. Published under the authority of the Lords Commissioners of the Admiralty.

The 'Flora of Tasmania' describes all the Plants, flowering and flowerless, of that Island, consisting of 2203 Species, collected by the Author and others. The Plates, of which there are 200, illustrate 412 Species.

HANDBOOK OF THE NEW ZEALAND FLORA;

a Systematic Description of the Native Plants of New Zealand, and the Chatham, Kermadec's, Lord Auckland's, Campbell's, and Macquarrie's Islands. By Dr. J. D. HOOKER, F.R.S. Demy 8vo, Part I., 475 pp., 16s. Published under the auspices of the Government of that colony.

[*Part II. in the Press.*

A compendious account of the plants of New Zealand and outlying islands, published under the authority of the Government of that colony. The present Part contains the Flowering Plants, Ferns, and Lycopods; the Second Part, containing the remaining Orders of *Cryptogamia*, or Flowerless Plants, with Index and Catalogues of Native Names and of Naturalized Plants, will appear shortly.

FLORA AUSTRALIENSIS;

a Description of the Plants of the Australian Territory. By GEORGE BENTHAM, F.R.S., President of the Linnean Society, assisted by FERDINAND MUELLER, F.R.S., Government Botanist, Melbourne, Victoria. Demy 8vo. Vol. I. 566 pp., and vol. II. 530 pp., 20s. each. Published under the auspices of the several Governments of Australia. [*Vol. III. nearly ready.*

Of this great undertaking, the present volumes, of more than a thousand closely-printed pages, comprise about one-fourth. The materials are derived not only from the vast collections of Australian plants brought to this country by various botanical travellers, and preserved in the herbaria of Kew and of the British Museum, including those hitherto unpublished of Banks and Solander, of Captain Cook's first Voyage, and of Brown in Flinders', but from the very extensive and more recently collected specimens preserved in the Government Herbarium of Melbourne, under the superintendence of Dr. Ferdinand Mueller. The descriptions are written in plain English, and are masterpieces of accuracy and clearness.

FLORA HONGKONGENSIS;

a Description of the Flowering Plants and Ferns of the Island of Hongkong. By GEORGE BENTHAM, P.L.S. With a Map of the Island. Demy 8vo, 550 pp., 16s. Published under the authority of Her Majesty's Secretary of State for the Colonies.

The Island of Hongkong, though occupying an area of scarcely thirty square miles, is characterized by an extraordinarily varied Flora, partaking, however, of that of South Continental China, of which comparatively little is known. The number of Species enumerated in the present volume is 1056, derived chiefly from materials collected by Mr. Hinds, Col. Champion, Dr. Hance, Dr. Harland, Mr. Wright, and Mr. Wilford.

FLORA OF THE BRITISH WEST INDIAN ISLANDS.

By Dr. GRISEBACH, F.L.S. Demy 8vo, 806 pp., 37s. 6d. Published under the auspices of the Secretary of State for the Colonies.

Containing complete systematic descriptions of the Flowering Plants and Ferns of the British West Indian Islands, accompanied by an elaborate index of reference, and a list of Colonial names.

FLORA VITIENSIS; a Description of the Plants of the Viti or Fiji Islands, with an Account of their History, Uses, and Properties. By Dr. BERTHOLD SEEMANN, F.L.S. Royal 4to, Parts I. to IV. each, 10 Coloured Plates, 15s. To be completed in 10 Parts.

This work owes its origin to the Government Mission to Viti, to which the uthor was attached as naturalist. In addition to the specimens collected, the author has investigated all the Polynesian collections of Plants brought to this country by various botanical explorers since the voyage of Captain Cook.

ILLUSTRATIONS OF THE NUEVA QUINOLOGIA OF PAVON, with Observations on the Barks described. By J. E. HOWARD, F.L.S. With 27 Coloured Plates by W. FITCH. Imperial folio, half-morocco, gilt edges, £6. 6s.

A superbly-coloured volume, illustrative of the most recent researches of Pavon and his associates among the Cinchona Barks of Peru, founded mainly on a manuscript and collection of specimens which were sold shortly before Pavon's death to a botanist of Madrid, from whom they passed into the hands of the author.

ILLUSTRATIONS OF SIKKIM-HIMALAYAN PLANTS, chiefly selected from Drawings made in Sikkim, under the superintendence of the late J. F. CATHCART, Esq., Bengal Civil Service. The Botanical Descriptions and Analyses by Dr. J. D. HOOKER, F.R.S. Imperial folio, 24 Coloured Plates and an Illuminated Title-page by W. FITCH, £5. 5s.

As an example of botanical drawing, colouring, and design, this work has never been surpassed. Only a few copies remain.

THE LONDON JOURNAL OF BOTANY. Original Papers by eminent Botanists, Letters from Botanical Travellers, etc. Vol. VII., completing the Series. Demy 8vo, 23 Plates, 30s.

JOURNAL OF BOTANY AND KEW MISCELLANY. Original Papers by eminent Botanists, Letters from Botanical Travellers, etc. Edited by Sir W. J. HOOKER, F.R.S. Vols. IV. to IX., Demy 8vo, 12 Plates, £1. 4s. A Complete Set of 9 vols., half-calf, *scarce*, £10. 16s.

ICONES PLANTARUM. Figures, with brief Descriptive Characters and Remarks, of New and Rare Plants, selected from the Author's Herbarium. By Sir W. J. HOOKER, F.R.S. New Series. Vol. V., Royal 8vo, 100 plates, 31s. 6d.

FERNS AND MOSSES.

THE BRITISH FERNS; or, Coloured Figures and Descriptions, with the needful Analyses of the Fructification and Venation, of the Ferns of Great Britain and Ireland, systematically arranged. By Sir W. J. HOOKER, F.R.S. Royal 8vo, 66 Plates, £2. 2s.

The British Ferns and their allies are illustrated in this work, from the pencil of Mr. FITCH. Each Species has a Plate to itself, so that there is ample room for the details, on a magnified scale, of Fructification and Venation. The whole are delicately coloured by hand. In the letterpress an interesting account is given with each species of its geographical distribution in other countries.

GARDEN FERNS; or, Coloured Figures and Descriptions, with the needful Analyses of the Fructification and Venation, of a Selection of Exotic Ferns, adapted for Cultivation in the Garden, Hothouse, and Conservatory. By Sir W. J. HOOKER, F.R.S. Royal 8vo, 64 Plates, £2. 2s.

A companion volume to the preceding, for the use of those who take an interest in the cultivation of some of the more beautiful and remarkable varieties of Exotic Ferns. Here also each Species has a Plate to itself, and the details of Fructification and Venation are given on a magnified scale, the Drawings being from the pencil of Mr. FITCH.

FILICES EXOTICÆ; or, Coloured Figures and Description of Exotic Ferns, chiefly of such as are cultivated in the Royal Gardens of Kew. By Sir W. J. HOOKER, F.R.S. Royal 4to, 100 Plates, £6. 11s.

One of the most superbly illustrated books of Foreign Ferns that has been hitherto produced. The Species are selected both on account of their beauty of form, singular structure, and their suitableness for cultivation.

FERNY COMBES; a Ramble after Ferns in the Glens and Valleys of Devonshire. By CHARLOTTTE CHANTER. *Second Edition.* Fcp. 8vo, 8 coloured plates by Fitch, and a Map of the County, 5s.

HANDBOOK OF BRITISH MOSSES, containing all that are known to be Natives of the British Isles. By the Rev. M. J. BERKELEY, M.A., F.L.S. Demy 8vo, pp. 360, 24 Coloured Plates, 21s.

A very complete Manual, comprising characters of all the species, with the circumstances of habitation of each; with special chapters on development and structure, propagation, fructification, geographical distribution, uses, and modes of collecting and preserving, followed by an extensive series of coloured illustrations, in which the essential portions of the plant are repeated, in every case on a magnified scale.

SEAWEEDS.

PHYCOLOGIA BRITANNICA; or, History of British Seaweeds, containing Coloured Figures, Generic and Specific Characters, Synonyms and Descriptions of all the Species of Algæ inhabiting the Shores of the British Islands. By Dr. W. H. HARVEY, F.R.S. Royal 8vo, 4 vols., 765 pp., 360 Coloured Plates, £6. 6s. Reissue in Monthly Parts, each 2s. 6d.

This work, originally published in 1851, at the price of £7. 10s., is still the standard work on the subject of which it treats. Each Species, excepting the minute ones, has a Plate to itself, with magnified portions of structure and fructification, the whole being printed in their natural colours, finished by hand.

SYNOPSIS OF BRITISH SEAWEEDS, compiled from Dr. Harvey's 'PHYCOLOGIA BRITANNICA.' Small 8vo, 220 pp., 5s.

A Descriptive Catalogue of all the British Seaweeds, condensed from the 'Phycologia Britannica.' It comprises the characters, synonyms, habitats, and general observations, forming an extremely useful pocket volume of reference.

PHYCOLOGIA AUSTRALICA; a History of Australian Seaweeds, comprising Coloured Figures and Descriptions of the more characteristic Marine Algæ of New South Wales, Victoria, Tasmania, South Australia and Western Australia, and a Synopsis of all known Australian Algæ. By Dr. HARVEY, F.R.S. Royal 8vo, 5 vols., 300 Coloured Plates, £7. 13s.

This beautiful work, the result of an arduous personal exploration of the shores of the Australian continent, is got up in the style of the 'Phycologia Britannica' by the same author. Each Species has a Plate to itself, with ample magnified delineations of fructification and structure, embodying a variety of most curious and remarkable forms.

NEREIS AUSTRALIS; or, Algæ of the Southern Ocean, being Figures and Descriptions of Marine Plants collected on the Shores of the Cape of Good Hope, the extra-tropical Australian Colonies, Tasmania, New Zealand, and the Antarctic Regions. By Dr. HARVEY, F.R.S. Imperial 8vo, 50 Coloured Plates, £2. 2s.

A selection of Fifty Species of remarkable forms of Seaweed, not included in the 'Phycologia Australica,' collected over a wider area.

FUNGI.

OUTLINES OF BRITISH FUNGOLOGY, containing Characters of above a Thousand Species of Fungi, and a Complete List of all that have been described as Natives of the British Isles. By the Rev. M. J. BERKELEY, M.A., F.L.S. Demy 8vo, 484 pp., 24 Coloured Plates, 30s.

Although entitled simply 'Outlines,' this is a good-sized volume, of nearly 500 pages, illustrated with more than 200 Figures of British Fungi, all carefully coloured by hand. Of above a thousand Species the characters are given, and a complete list of the names of all the rest.

THE ESCULENT FUNGUSES OF ENGLAND. Containing an Account of their Classical History, Uses, Characters, Development, Structure, Nutritions Properties, Modes of Cooking and Preserving, etc. By C. D. BADHAM, M.D. Second Edition. Edited by F. CURREY, F.R.S. Demy 8vo, 152 pp., 12 Coloured Plates, 12s.

A lively classical treatise, written with considerable epigrammatic humour, with the view of showing that we have upwards of 30 Species of Fungi abounding in our woods capable of affording nutritious and savoury food, but which, from ignorance or prejudice, are left to perish ungathered. "I have indeed grieved," says the Author, "when reflecting on the straitened condition of the lower orders, to see pounds of extempore beefsteaks growing on our oaks, in the shape of *Fistulina hepatica*; Puff-balls, which some have not inaptly compared to sweetbread; *Hydna*, as good as oysters; and *Agaricus deliciosus*, reminding us of tender lamb-kidney." Superior coloured Figures of the Species are given from the pencil of Mr. Fitch.

ILLUSTRATIONS OF BRITISH MYCOLOGY, comprising Figures and Descriptions of the Funguses of interest and novelty indigenous to Britain. By Mrs. T. J. HUSSEY. Royal 4to; First Series, 90 Coloured Plates, £7. 12s. 6d.; Second Series, 50 Coloured Plates, £4. 10s.

This beautifully-illustrated work is the production of a lady who, being an accomplished artist, occupied the leisure of many years in accumulating a portfolio of exquisite drawings of the more attractive forms and varieties of British Fungi. The publication was brought to an end with the 140th Plate by her sudden decease. The Figures are mostly of the natural size, carefully coloured by hand.

SHELLS AND MOLLUSKS.

ELEMENTS OF CONCHOLOGY; an Introduction to the Natural History of Shells, and of the Animals which form them. By LOVELL REEVE, F.L.S. Royal 8vo, 2 vols., 478 pp., 62 Coloured Plates, £2. 16s.

Intended as a guide to the collector of shells in arranging and naming his specimens, while at the same time inducing him to study them with reference to their once living existence, geographical distribution, and habits. Forty-six of the plates are devoted to the illustration of the genera of shells, and sixteen to shells with the living animal, all beautifully coloured by hand.

THE LAND AND FRESHWATER MOLLUSKS indigenous to, or naturalized in, the British Isles. By LOVELL REEVE, F.L.S. Crown 8vo, 295 pp., Map, and 160 Wood-Engravings, 10s. 6d.

A complete history of the British Land and Freshwater Shells, and of the Animals which form them, illustrated by Wood-Engravings of all the Species. Other features of the work are an Analytical Key, showing at a glance the natural groups of families and genera, copious Tables and a Map illustrative of geographical distribution and habits, and a chapter on the Distribution and Origin of Species.

CONCHOLOGIA ICONICA; or, Figures and Descriptions of the Shells of Mollusks, with remarks on their Affinities, Synonymy, and Geographical Distribution. By LOVELL REEVE, F.L.S. Demy 4to, published monthly in Parts, 8 Plates, carefully coloured by hand, 10s.

Of this work, comprising illustrations of Shells of the natural size, nearly 2000 Plates are published, but the plan of publication admits of the collector purchasing it at his option in portions, each of which is complete in itself. Each genus, as the work progresses, is issued separately, with Title and Index; and an Alphabetical List of the published genera, with the prices annexed, may be procured of the publishers on application. The system of nomenclature adopted is that of Lamarck, modified to meet the exigencies of later discoveries. With the name of each species is given a summary of its leading specific characters in Latin and English; then the authority for the name is quoted, accompanied by a reference to its original description; and next in order are its Synonyms. The habitat of the species is next given, accompanied, where possible, by particulars of soil, depth, or vegetation. Finally, a few general remarks are offered, calling attention to the most obvious distinguishing peculiarities of the species, with criticisms, where necessary, on the views of other writers. At the commencement of the genus some notice is taken of the animal, and the habitats of the species are worked up into a general summary of the geographical distribution of the genus.

CONCHOLOGIA ICONICA IN MONOGRAPHS.

Genera.	Plates.	£.	s.	d.	Genera.	Plates.	£.	s.	d.
Achatina	23	1	9	0	Io	3	0	4	0
Achatinella	6	0	8	0	Isocardia	1	0	1	6
Adamsiella	2	0	3	0	Leptopoma	8	0	10	6
Amphidesma	7	0	9	0	Lingula	2	0	3	0
Ampullaria	28	1	15	6	Lithodomus	5	0	6	6
Anastoma	1	0	1	6	Littorina	18	1	3	0
Anatina	4	0	5	6	Lucina	11	0	14	0
Ancillaria	12	0	15	6	Lutraria	5	0	6	6
Anculotus	6	0	8	0	Mactra	21	1	6	6
Anomia	8	0	10	6	Malleus	3	0	4	0
Arca	17	1	1	6	Mangelia	8	0	10	6
Argonauta	4	0	5	6	Marginella	27	1	14	6
Artemis	10	0	13	0	Melania	59	3	14	6
Aspergillum	4	0	5	6	Melanopsis	3	0	4	0
Avicula	18	1	3	0	Melatoma	3	0	4	0
Buccinum	14	0	18	0	Meroe	3	0	4	0
Bulimus	89	5	12	0	Mesalia & Eglisia	1	0	1	6
Bullia	4	0	5	6	Mesodesma	4	0	5	6
Calyptræa	8	0	10	6	Meta	1	0	1	6
Cancellaria	18	1	3	0	Mitra	39	2	9	6
Capsa	1	0	1	6	Modiola	11	0	14	0
Capsella	2	0	3	0	Monoceros	4	0	5	6
Cardita	9	0	11	6	Murex	37	2	7	0
Cardium	22	1	8	0	Myadora	1	0	1	6
Carinaria	1	0	1	6	Myochama	1	0	1	6
Cassidaria	1	0	1	6	Mytilus	11	0	14	0
Cassis	12	0	15	6	Nassa	29	1	17	0
Cerithium	20	1	5	6	Natica	30	1	18	0
Chama	9	0	11	6	Nautilus	6	0	8	0
Chamostrea	1	0	1	6	Navicella & Latia	8	0	10	6
Chiton	33	2	2	0	Nerita	19	1	4	0
Chitonellus	1	0	1	6	Neritina	37	2	7	0
Chondropoma	11	0	14	0	Oliva	30	1	18	0
Circe	10	0	13	0	Oniscia	1	0	1	6
Columbella	37	2	7	0	Orbicula	1	0	1	6
Concholepas	2	0	3	0	Ovulum	14	0	18	0
Conus	56	3	11	0	Paludina	11	0	14	0
Corbula	5	0	6	6	Paludomus	3	0	4	0
Crania	1	0	1	6	Partula	4	0	5	6
Crassatella	3	0	4	0	Patella	42	2	13	0
Crenatula	2	0	3	0	Pecten	35	2	4	6
Crepidula	5	0	6	6	Pectunculus	9	0	11	6
Crucibulum	7	0	9	0	Pedum	1	0	1	6
Cyclophorus	20	1	5	6	Perna	6	0	8	0
Cyclostoma	23	1	9	0	Phasianella	6	0	8	0
Cyclotus	9	0	11	6	Phorus	3	0	4	0
Cymbium	26	1	13	0	Pinna	34	2	3	0
Cypræa	27	1	14	6	Pirena	2	0	3	0
Cypricardia	2	0	3	0	Placunanomia	3	0	4	0
Cytherea	10	0	13	0	Pleurotoma	40	2	10	6
Delphinula	5	0	6	6	Psammobia	8	0	10	6
Dione	12	0	15	6	Psammotella	1	0	1	6
Dolium	8	0	10	6	Pterocera	6	0	8	0
Donax	9	0	11	6	Pterocyclos	5	0	6	6
Eburna	1	0	1	6	Purpura	13	0	16	6
Erato	3	0	4	0	Pyramidella	6	0	8	0
Eulima	6	0	8	0	Pyrula	9	0	11	6
Fasciolaria	7	0	9	0	Ranella	8	0	10	6
Ficula	1	0	1	6	Ricinula	6	0	8	0
Fissurella	16	1	0	6	Rostellaria	3	0	4	6
Fusus	21	1	6	6	Sanguinolaria	1	0	1	6
Glauconome	1	0	1	6	Scarabus	3	0	4	0
Halia	1	0	1	6	Sigaretus	5	0	6	6
Haliotis	17	1	1	6	Simpulopsis	2	0	3	0
Harpa	4	0	5	6	Siphonaria	7	0	9	0
Helix	210	13	5	0	Solarium	3	0	4	0
Hemipecten	1	0	1	6	Soletellina	4	0	5	6
Hemisinus	6	0	8	0	Spondylus	18	1	3	6
Hinnites	1	0	1	6	Strombus	19	1	4	0
Hippopus	1	0	1	6	Struthiolaria	1	0	1	6
Ianthina	5	0	6	6	Tapes	13	0	16	6

CONCHOLOGIA SYSTEMATICA; or, Complete System of
Conchology. By LOVELL REEVE, F.L.S. Demy 4to, 2 vols. pp. 537,
300 Plates, £8. 8s. coloured.

Of this work only a few copies remain. It is a useful companion to the collector of shells, on account of the very large number of specimens figured, as many as six plates being devoted in some instances to the illustration of a single genus.

INSECTS.

CURTIS' BRITISH ENTOMOLOGY. Illustrations and Descriptions of the Genera of Insects found in Great Britain and Ireland, containing Coloured Figures, from nature, of the most rare and beautiful species, and, in many instances, upon the plants on which they are found. Royal 8vo, 8 vols., 770 Plates, coloured, £21.

Or in separate Monographs.

Orders.	Plates.	£	s.	d.	Orders.	Plates.	£	s.	d.
APHANIPTERA	2	0	2	0	HYMENOPTERA	125	4	0	0
COLEOPTERA	256	8	0	0	LEPIDOPTERA	193	6	0	0
DERMAPTERA	1	0	1	0	NEUROPTERA	13	0	9	0
DICTYOPTERA	1	0	1	0	OMALOPTERA	6	0	4	6
DIPTERA	103	3	5	0	ORTHOPTERA	5	0	4	0
HEMIPTERA	32	1	1	0	STREPSIPTERA	3	0	2	6
HOMOPTERA	21	0	14	0	TRICHOPTERA	9	0	6	6

'Curtis' Entomology,' which Cuvier pronounced to have "reached the ultimatum of perfection," is still the standard work on the Genera of British Insects. The Figures executed by the author himself, with wonderful minuteness and accuracy, have never been surpassed, even if equalled. The price at which the work was originally published was £43. 16s.

INSECTA BRITANNICA; Vols. II. and III., Diptera. By
FRANCIS WALKER, F.L.S. 8vo, each, with 10 plates, 25s.

TRAVELS.

THREE CITIES IN RUSSIA. By Professor C. PIAZZI SMYTH, F.R.S. Post 8vo, 2 Vols., 1016 pp. Maps and Wood-Engravings, 26s.

The narrative of a tour made in the summer of 1859 by the Astronomer Royal of Scotland, to the cities of St. Petersburg, Moscow, and Novgorod.

THE GATE OF THE PACIFIC. By Commander BEDFORD PIM, R.N. Demy 8vo, 430 pp., with 7 Maps and 8 Tinted Chromo-Lithographs, 18s.

A spirited narrative of Commander Pim's explorations in Central America, made with the view of establishing a new overland route from the Atlantic to the Pacific Oceans, through English enterprise, by way of Nicaragua.

TRAVELS ON THE AMAZON AND RIO NEGRO; with an Account of the Native Tribes, and Observations on the Climate, Geology, and Natural History of the Amazon Valley. By ALFRED R. WALLACE. Demy 8vo, 541 pp., with Map and Tinted Frontispiece, 18s.

A lively narrative of travels in one of the most interesting districts of the Southern Hemisphere, accompanied by Remarks on the Vocabularies of the Languages, by Dr. R. G. LATHAM.

WESTERN HIMALAYA AND TIBET; a Narrative of a Journey through the Mountains of Northern India, during the Years 1847-1848. By Dr. THOMSON, F.R.S. Demy 8vo, 500 pp., with Map and Tinted Frontispiece, 15s.

A summary of the physical features, chiefly botanical and geological, of the country travelled over in a mission undertaken for the Indian Government, from Simla across the Himalayan Mountains into Tibet, and to the summit of the Karakoram Mountains; including also an excellent description of Kashmir.

TRAVELS IN THE INTERIOR OF BRAZIL, principally through the Northern Provinces and the Gold and Diamond Districts, during the years 1836–1841. By Dr. GEORGE GARDNER, F.L.S. Second Edition. Demy 8vo, 428 pp., with Map and Tinted Frontispiece, 12s.

The narrative of an arduous journey, undertaken by an enthusiastic naturalist, through Brazil Proper, Bahia, Maranham, and Pernambuco, written in a lively style, with glowing descriptions of the grandeur of the vegetation.

ANTIQUARIAN.

MAN'S AGE IN THE WORLD ACCORDING TO
HOLY SCRIPTURE AND SCIENCE. By an Essex Rector. Demy 8vo, 264 pp., 8s. 6d.

 The Author, recognizing the established facts and inevitable deductions of Science, and believing all attempts to reconcile them with the commonly received, but erroneous, literal interpretation of Scripture, not only futile, but detrimental to the cause of Truth, seeks an interpretation of the Sacred Writings on general principles, consistent alike with their authenticity, when rightly understood, and with the exigencies of Science. He treats in successive chapters of The Flint Weapons of the Drift,—The Creation,—The Paradisiacal State,—The Genealogies,—The Deluge,—Babel and the Dispersion; and adds an Appendix of valuable information from various sources.

THE ANTIQUITY OF MAN. An Examination of Sir
Charles Lyell's recent Work. By S. R. Pattison, F.G.S. Second Edition. 8vo, 1s.

HORÆ FERALES; or, Studies in the Archæology of the
Northern Nations. By the late John M. Kemble, M.A. Edited by Dr. R. G. Latham, F.R.S., and A. W. Franks, M.A. Royal 4to, 263 pp., 34 Plates, many coloured, £3. 3s.

A MANUAL OF BRITISH ARCHÆOLOGY. By
Charles Boutell, M.A. Royal 16mo, 398 pp., 20 coloured plates, 10s. 6d.

 A treatise on general subjects of antiquity, written especially for the student of archæology, as a preparation for more elaborate works. Architecture, Sepulchral Monuments, Heraldry, Seals, Coins, Illuminated Manuscripts and Inscriptions, Arms and Armour, Costume and Personal Ornaments, Pottery, Porcelain and Glass, Clocks, Locks, Carvings, Mosaics, Embroidery, etc., are treated of in succession, the whole being illustrated by 20 attractive Plates of Coloured Figures of the various objects.

SHAKESPEARE'S SONNETS, Facsimile, by Photo-Zinco-
graphy, of the First Printed edition of 1609. From the Copy in the Library of Bridgewater House, by permission of the Right Hon. the Earl of Ellesmere. 10s. 6d.

MISCELLANEOUS.

MANUAL OF CHEMICAL ANALYSIS, Qualitative and
Quantitative; for the Use of Students. By Dr. HENRY M. NOAD, F.R.S. Crown 8vo, pp. 663, 109 Wood Engravings, 16s. Or, separately, Part I., 'QUALITATIVE,' 6s.; Part II., 'QUANTITATIVE,' 10s. 6d.

A Copiously-illustrated, Useful, Practical Manual of Chemical Analysis, prepared for the Use of Students by the Lecturer on Chemistry at St. George's Hospital. The illustrations consist of a series of highly-finished Wood-Engravings, chiefly of the most approved forms and varieties of apparatus.

DICTIONARY OF NATURAL HISTORY TERMS, with
their Derivatives, including the various Orders, Genera, and Species. By DAVID H. M'NICOLL, M.D. Crown 8vo, 584 pp., 12s. 6d.

An attempt to furnish what has long been a desideratum in natural history,— a dictionary of technical terms, with their meanings and derivatives.

PHOSPHORESCENCE; or, the Emission of Light by Minerals,
Plants, and Animals. By Dr. T. L. PHIPSON, F.C.S. Small 8vo, 225 pp., 30 Wood Engravings and Coloured Frontispiece, 5s.

An interesting summary of the various phosphoric phenomena that have been observed in nature,—in the mineral, in the vegetable, and in the animal world.

A SURVEY OF THE EARLY GEOGRAPHY OF
WESTERN EUROPE, as connected with the First Inhabitants of Britain, their Origin, Language, Religious Rites, and Edifices. By HENRY LAWES LONG, Esq. 8vo, 6s.

THE ZOOLOGY OF THE VOYAGE OF H.M.S. SA-
MARANG, under the command of Captain Sir Edward Belcher, C.B., during the Years 1843–46. By Professor OWEN, Dr. J. E. GRAY, Sir J. RICHARDSON, A. ADAMS, L. REEVE, and A. WHITE. Edited by ARTHUR ADAMS, F.L.S. Royal 4to, 257 pp., 55 Plates, mostly coloured, £3. 10s.

In this work, illustrative of the new species of animals collected during the surveying expedition of H.M.S. Samarang in the Eastern Seas in the years 1843–1846, there are 7 Plates of Quadrupeds, 1 of Reptiles, 10 of Fishes, 24 of Mollusca and Shells, and 13 of Crustacea. The Mollusca, which are particularly interesting, include the anatomy of *Spirula* by Professor Owen, and a number of beautiful Figures of the living animals by Mr. Arthur Adams.

THE GEOLOGIST. A Magazine of Geology, Palæontology, and Mineralogy. Illustrated with highly finished Wood-Engravings. Edited by S. J. MACKIE, F.G.S., F.S.A. Vols. V. and VI., each, with numerous Wood-Engravings, 18s. Vol. VII., 9s.

OUTLINES OF ELEMENTARY BOTANY, as Introductory to Local Floras. By GEORGE BENTHAM, F.R.S., President of the Linnean Society. Demy 8vo, pp. 45, 2s. 6d.

ON THE FLORA OF AUSTRALIA, its Origin, Affinities, and Distribution; being an Introductory Essay to the 'Flora of Tasmania.' By Dr. J. D. HOOKER, F.R.S. 128 pp., quarto, 10s.

CRYPTOGAMIA ANTARCTICA; or, Cryptogamic Plants of the Antarctic Islands. Issued separately. In One Volume, quarto, £4. 4s. coloured; £2. 17s. plain.

GUIDE TO COOL-ORCHID GROWING. By JAMES BATEMAN, Esq., F.R.S., Author of 'The Orchidaceæ of Mexico and Guatemala.' Woodcuts, 1s.

A TREATISE ON THE GROWTH AND FUTURE TREATMENT OF TIMBER TREES. By G. W. NEWTON, of Ollersett, J.P. Half-bound calf, 10s. 6d.

PARKS AND PLEASURE GROUNDS; or, Practical Notes on Country Residences, Villas, Public Parks, and Gardens. By CHARLES H. J. SMITH, Landscape Gardener. Crown 8vo, 6s.

LITERARY PAPERS ON SCIENTIFIC SUBJECTS. By the late Professor EDWARD FORBES, F.R.S., selected from his Writings in the 'Literary Gazette.' With a Portrait and Memoir. Small 8vo, 6s.

THE PLANETARY AND STELLAR UNIVERSE. A Series of Lectures. With Illustrations. By R. J. MANN. 12mo, 5s.

THE STEREOSCOPIC MAGAZINE. A Gallery for the Stereoscope of Landscape Scenery, Architecture, Antiquities, Natural History, Rustic Character, etc. With Descriptions. 5 vols., each complete in itself and containing 50 Stereographs, £2. 2s.

THE CONWAY. Narrative of a Walking Tour in North Wales; accompanied by Descriptive and Historical Notes. By J. B. DAVIDSON, Esq., M.A. Extra gilt, 20 stereographs of Welsh Scenery, 21s.

THE ARTIFICIAL PRODUCTION OF FISH. By PISCARIUS. Third Edition. 1s.

NEWEST WORKS.

A SECOND CENTURY OF ORCHIDACEOUS PLANTS, selected from the subjects published in Curtis' 'Botanical Magazine' since the issue of the 'First Century.' Edited by JAMES BATEMAN, Esq., F.R.S. Parts I. and II., each with 10 Coloured Plates, 10s. 6d., now ready.
[*Part III. nearly ready.*

During the fifteen years that have elapsed since the publication of the 'Century of Orchidaceous Plants,' now out of print, the 'Botanical Magazine' has been the means of introducing to the public nearly two hundred of this favourite tribe of plants not hitherto described and figured, or very imperfectly so. It is intended from these to select "a Second Century," and the descriptions, written at the time of publication by Sir W. J. Hooker, will be edited, and in many cases re-written, agreeably with the present more advanced state of our knowledge and experience in the cultivation of Orchidaceous plants, by Mr. Bateman, the acknowledged successor of Dr. Lindley as the leading authority in this department of botany and horticulture. The size of the work is a handsome royal quarto, and it is proposed to issue the hundred plates in ten Parts, each containing ten plates, carefully coloured by hand, price 10s. 6d.

THE BEWICK COLLECTOR. A Descriptive Catalogue of the Works of THOMAS and JOHN BEWICK, including Cuts, in various states, for Books and Pamphlets, Private Gentlemen, Public Companies, Exhibitions, Races, Newspapers, Shop Cards, Invoice Heads, Bar Bills, Coal Certificates, Broadsides, and other miscellaneous purposes, and Wood Blocks. With an Appendix of Portraits, Autographs, Works of Pupils, etc. The whole described from the Originals contained in the Largest and most Perfect Collection ever formed, and illustrated with a Hundred and Twelve Cuts from Bewick's own Blocks. By the Rev. THOMAS HUGO, M.A., F.S.A., the Possessor of the Collection. Demy 8vo, pp. 562, price 21s.; imperial 8vo (limited to 100 copies), with a fine Steel Engraving of Thomas Bewick, £2. 2s.

Commencement of a New Series of Natural History for Beginners.

BRITISH BEETLES; a Familiar Introduction to the study of our Indigenous COLEOPTERA. By E. C. RYE. Crown 8vo, 16 Coloured Steel Plates, comprising Figures of nearly 100 Species, engraved from Natural Specimens, expressly for the work, by E. W. ROBINSON, and 11 Wood-Engravings of Dissections by the Author, 10s. 6d. [Ready.

BRITISH SPIDERS; a Familiar Introduction to the study of our Native ARACHNIDA. By E. F. STAVELEY. Crown 8vo, 16 Coloured Plates and Wood-Engravings, 10s. 6d. [Nearly ready.

BRITISH BEES; a Familiar Introduction to the study of our Native Bees. By W. E. SCHUCKARD. Crown 8vo, 16 Coloured Plates, and Wood-Engravings, 10s. 6d. [Ready.

BRITISH BUTTERFLIES AND MOTHS; a Familiar Introduction to the study of our Native LEPIDOPTERA. By H. T. STAINTON. Crown 8vo, 16 Coloured Plates, and Wood-Engravings, 10s. 6d. [In preparation.

BRITISH FERNS: a Familiar Introduction to the Study of our Native Ferns and their Allies. By MARGARET PLUES. Crown 8vo, 16 Coloured Plates, and Wood-Engravings, 10s. 6d. [Ready.

BRITISH SEAWEEDS; a Familiar Introduction to the study of our Native Marine ALGÆ. By S. O. GRAY. Crown 8vo, 16 Coloured Plates, and Wood-Engravings, 10s. 6d. [In preparation.

*** A good introductory series of books on British Natural History for the use of students and amateurs is still a *desideratum*. Those at present in use have been too much compiled from antiquated sources; while the figures, copied in many instances from sources equally antiquated, are far from accurate, the colouring of them having become degenerated through the adoption, for the sake of cheapness, of mechanical processes.

The present series will be entirely the result of original research carried to its most advanced point; and the figures, which will be chiefly engraved on steel, by the artist most highly renowned in each department for his technical knowledge of the subjects, will in all cases be drawn from actual specimens, and coloured separately by hand.

LONDON:
LOVELL REEVE & CO., 5, HENRIETTA STREET, COVENT GARDEN.

www.ingramcontent.com/pod-product-compliance
Lightning Source LLC
Chambersburg PA
CBHW022106300426
44117CB00007B/603